Olympic Villages and Urban Development

Olympic Villages and Urban Development

Analysis of Spatial Models and Geographic Transformations

Valerio della Sala

PETER LANG

Oxford - Berlin - Bruxelles - Chennai - Lausanne - New York

Bibliographic information published by the Deutsche Nationalbibliothek.
The German National Library lists this publication in the German National Bibliography;
detailed bibliographic data is available on the Internet at http://dnb.d-nb.de.

A catalogue record for this book is available from the British Library.

Library of Congress Cataloging-in-Publication Data

Names: Della Sala, Valerio, 1990- author.
Title: Olympic villages and urban development : analysis of spatial models
    and geographic transformations / Valerio della Sala.
Description: Oxford ; New York : Peter Lang, [2025] | Includes
    bibliographical references and index.
Identifiers: LCCN 2024031347 (print) | LCCN 2024031348 (ebook) | ISBN
    9781803742946 (paperback) | ISBN 9781803742953 (ebook) | ISBN
    9781803742960 (epub)
Subjects: LCSH: Olympic host city selection—Social aspects. | Olympic host
    city selection—Economic aspects. | City planning. | Cities and
    towns—Growth—Management. | Olympics—Planning.
Classification: LCC GV721.5 .D4475 2025 (print) | LCC GV721.5 (ebook) |
    DDC 307.1/4—dc23/eng/20240816
LC record available at https://lccn.loc.gov/2024031347
LC ebook record available at https://lccn.loc.gov/2024031348

The study was supported by the Olympic Studies Centre of the Autonomous University
of Barcelona, Interdepartmental Research Centre in Urban Studies and Events (OMERO)
of the University of Turin and, Sport Research Institute (IRE-UAB) of the Universitat
Autonoma de Barcelona.

Cover image: Valerio della Sala, personal archive.
Cover design by Peter Lang Group AG

ISBN 978-1-80374-294-6 (print)
ISBN 978-1-80374-295-3 (ePDF)
ISBN 978-1-80374-296-0 (ePub)
DOI 10.3726/b22095

© 2025 Peter Lang Group AG, Lausanne
Published by Peter Lang Ltd, Oxford, United Kingdom
info@peterlang.com - www.peterlang.com

This publication has been peer reviewed.

Contents

Brief Synopsis

One of the critical elements of my work is the relationship between urban transformation and its future implications in the territory, promoted by Olympic urban planning. Not being thoroughly analysed at the academic level and observing the lack of previous and comparative studies on the subject, it is stated that, given the complexity of the subject, a theoretical framework has been formulated that pursues the following research objectives. Observing the territorial impact of the Olympic Villages in the city, introducing new organisation models, and financing the Olympic event are considered essential in analysing the territorialisation and deterritorialisation processes linked to the territory and its transformation throughout the weather. This assumption made by Turco (1988) will be necessary for understanding the impact of the Olympic Village on the regional territory. The creation of new structures will cause territorial changes and recent incidents in the Olympic cities that will be difficult to remove shortly. After having established these concepts about the Olympic games and, therefore, about the territorial impact of the Olympic Villages, to analyse the effect over time, the study will examine the planning of the Olympic Village over time in all its development phases. Later, the research will explore the Olympic Games according to their specific function as urban catalysts provided by Muñoz (1996). Based on this concept of territorial link and affirming the importance of Olympic urban planning in the candidate cities, other contributions are presented to the infrastructural system and the event's organisation in the Olympic cities. Thanks to this contribution, we can affirm the importance of Olympic urbanism as a catalyst for the future urban growth and development of the candidate cities. At the same time, the Olympic Village can generate new specific strategies for developing the territory and using space. The processes of territorialisation and regionalisation are fundamental to understanding the impact of the Olympic Village on the territory. Creating new structures will always entail territorial changes and further incidences in the internal connections of the city and the territory.

Meanwhile, the reports provided by the UN for the sustainable development of cities and mega sports events will be considered. The analysis will look at all documentation produced since the Agenda 2020 of Rio de Janeiro in 1992 and the new guidelines for sustainable development in 2030. The importance of studying this issue through a new concept of observation about the different phases of the project and how it will change over time is affirmed. I want to underline the importance of my analytical study on the planning of the Olympic Village and its territorial impact. By proposing and defining a new development model for the candidate cities, it will be possible to rethink the Olympic Village as a temporary catalytic instrument. The research (see della Sala, 2022) proposes an innovative analysis of Olympic urban planning in general, and of Olympic Villages in particular, which will both consolidate the field of study that Olympic urban planning represents and offer a helpful document about the future construction of new Olympic Villages in the coming decades. The proposal aims to disseminate knowledge and propose new models of Olympic Village management in the post-Olympic period.

Figures

Tables

Acknowledgements

The following text was made possible thanks to a doctoral research project in international co-tutorship between the Autonomous University of Barcelona, the Polytechnic University of Turin and the University of Turin. Over time, the following study has received support from many specialists who continue to support me in my research work.

First, I thank Professor Francesc Muñoz and Professor Egidio Dansero, sources of inspiration and daily encouragement. Likewise, I would like to thank all the Centre for Olympic Studies members at the Autonomous University of Barcelona, the Institute of Sport Research (IRE-UAB) and the OMERO Interdepartmental Centre for Urban Studies at the University of Turin.

Collaborating with the following research centres has allowed me to produce different publications that are fundamental to my research experience.

Finally, I would like to thank my family, my father Roberto and my mother Katia for always having supported and sustained me. Special thanks also go to my partner, Chiara, who has always tried to help me with my choices from the very first moment. I want to emphasise that this book is also the result of extensive research in the Piedmont region, made possible thanks to the suggestions of some very special people in the OMERO group.

Grazie!

FRANCESC MUÑOZ

Foreword: Decoding "Olympic Urbanism" – The Olympic Villages of Yesterday and Tomorrow

In the course of the twentieth century, the relationship between cities and major events has become a commonplace that illustrates and explains a great variety of urban experiences of intense transformation not only of the physical urban space but also, above all, with regard to the image of the city. In this sense, the organisation of the Olympic Games occupies a predominant place in this process of tangible and intangible mutation of the city which, despite its obviously occasional and exceptional character, has represented an important urban, social and cultural footprint in quite a few Olympic experiences.

At the First Congress on Olympic Cities, held in Lausanne in 1996, the foundations were laid to start considering the urban transformation specific to the organisation of the Olympic Games as a special type of urbanism. An urbanism that, Olympiad after Olympiad, always showed itself as a mirror image of the ambitions and challenges of the organising cities, many times not translated from the urban planning dimension and the exercise of conventional urban management. Since then, research on this peculiar type of urban transformation has underlined the primary role of the "Olympic Village" as a nodal element in these processes of morphological and functional change in the urban space.

In this sense, the work of Valerio della Sala, not only gives continuity to this historical trajectory of Olympic urban studies but manages to make available a certainly definitive research, which not only illustrates and explains the complexity of the Olympic village as a minimum cell of Olympic urban planning, but also allows us to understand various territorial cases as a result of a very ambitious analysis that not only considers the Olympic villages of the Summer Olympic Games but also those built for the purpose of the Winter Olympics. In this sense, we are faced with a certainly total contribution regarding the object of study.

The first and main contribution of this book consists, therefore, in the affirmation and demonstration of the existence of an "Olympic Urbanism". A concrete and specific type of urban intervention with its own and differentiating characteristics, which have to do with three key issues that the author explores in great detail throughout the different chapters of the book and which constitute other major contributions of this work:

Firstly, what the author points out as the process of "territorialisation" of the Olympic village and which is explained on the basis of a thorough analysis of all the experiences developed throughout the history of the Modern Olympic Games, during the twentieth and twenty-first centuries. In this sense, the ambition of the study carried out is truly remarkable, with an exhaustive description of spatial occupation models, built-up areas or public space endowments, but, above all, it must be highlighted the attempt, certainly successfully resolved, to propose a general typology of all interventions from an urban and also regional perspective.

In this way, the taxonomy that differentiates the Olympic villages based on their models of territorial implementation – Monocentric; Satellite; Polycentric; Cluster; Peripheral; Metropolitan –, is not only correct and suggestive, but also allows us to distinguish patterns and generalisations in the analysis of an object of study that has often been defined by its ambiguity, dispersion or extreme diversity, so that it has ended up abusing, not infrequently, a rather descriptive and not very explanatory approach to the phenomenon. This work, therefore, contributes to the study of Olympic urbanism from the very opposite.

The second major contribution of this book consists in the emphasis placed on the legacy of the Olympic Games, defined from the built footprint that represents the Olympic village in the urban context. In this sense, the author proposes a true deconstruction of the idea of the "Olympic legacy", so often limited to some specific elements within the globality of strategies that make up the organisation of the Games, such as the cultural, communicative dimension or the linked to the construction of the collective consensus regarding the celebration of the mega-event.

Without at all denying the interest of these perspectives, the work of Valerio della Sala succeeds in elevating the category of legacy to a differential sphere in two different senses: firstly, by understanding legacy from a much more global perspective, regarding the characterisation of the Olympic Games as a model of urban mega-event. Secondly, raising the idea of legacy directly linked to the urban planning process itself, both in terms of the short term – the actual construction of the Olympic village while incorporating the perspective of legacy – and in the long run – the process of dialogue between the new Olympic neighbourhood and the rest of the city's own urban dynamics that will take place after the Games.

Based in these considerations, the definition of the Olympic village as a catalyst of urban change processes is very remarkable because it projects Olympic urban planning beyond the time of the Games in the strict sense and suggests its influence and footprint in terms of the future physical evolution but also in relation to the new collective image, locally and globally, of the city after the mega-event.

The third major contribution of the book lies in the ambition to determine, in the light of the research carried out and the experiences of Olympic villages built during the last 100 years, a model of intervention specific to Olympic urbanism in terms of the Games that will be organised in the future. In this sense, the author presents a series of keys in order to define how to approach the Olympic villages of the future Games, among which the idea of legacy and the necessary sustainability occupy a primary place.

Finally, and above other considerations, this book starts from those first seminal researches carried out in the light of that First Symposium of Olympic Villages, now thirty years ago, in order to establish clearly and on the basis of extensive research and exhaustive, a true new field of study and disciplinary research, defined in the most specific and solid way: Olympic urbanism.

The timely publication of this volume just as the Centenary Games are being celebrated in Paris in 2024, also allows to perfectly frame the reflections proposed by the author both regarding the past, while reflecting on the lessons learned from of those Paris Games in 1924, as regards, above all, the future of Olympic urbanism and the Olympic villages that will shape it in the twenty-first century.

Introduction

Mega-events are considered a particular phenomenon that has seen a significant increase in interest in recent years, fuelling the proliferation of smaller events on a local or continental scale. The large metropolises remain interested in the competition and future allocation of a global event that can be a catalyst for other meanings in the socio-cultural context of the host cities.

London, Paris, New York, Moscow, Boston, Madrid, Melbourne, Rome, Milan, Beijing, Los Angeles and Barcelona are just some of the large metropolises that continue to compete for the Olympic event in their own countries. The interest in hosting an event of this size can have different meanings and outcomes in each context. Over the years, we have observed various experiences that allow us to state that the possibilities of reconstructing urban functions, improving and expanding global networks, and consequently modifying and proposing to the world a new image of the place on a symbolic and material level, are the main catalysing elements that we will observe in the study.

The approach chosen in this work is territorial, starting with the first question posed to host communities before a significant event: What urban functions have Olympic Villages taken on throughout history, considering their use in the post-Olympic period?

The following question can be adequately answered by looking at the different models of Olympic Villages developed over time.

Therefore, can the Olympic Village be considered a catalyst for urban expansion and/or transformation in the planning of host cities?

From a local point of view, the event is a catalyst for new processes that can allow the city and the territory to temporarily take advantage of material and immaterial resources that, planned through strategic plans, can be reused and exploited in the post-Olympic period, thus becoming a new model of local development that the territory wants to implement in the medium and long term.

All the works and resources constitute a territorial legacy that can be transformed into a positive experience that becomes part of the city's available heritage and territorial resources. On the other hand, the Olympic experience can represent a harmful and disruptive element for the evolution of phenomena such as territorial fragmentation, segregation, gentrification and other phenomena that can be a problem and a cost for society. The definition of adequate and long-term planning can enable the implementation of projects that meet the needs of citizens and the specific local context.

One of the book's objectives was to identify and explore the Olympic Village experiences that constitute the most important and impactful Olympic heritage in the host cities. Therefore, the interpretation of the principal territorial transformations of the Olympic cities will focus on the evolution of both editions (summer and winter) to observe the different phases that characterised the evolution of the Olympic event. Subsequently, after watching the different evolutionary stages, the Olympic Villages built during the twentieth and twenty-first centuries will be introduced. Therefore, the following contribution will consider the Olympic territory as a phenomenon of extraordinary territory production that must be compared with the ordinary output of the territory. Furthermore, considering the event as a remarkable territory production allows us to observe different institutional and non-institutional rules, processes and obligations. The following text will assume all the resources, relationships and responsibilities provided by the Olympic charter.

The Olympic event permanently transforms and complexifies the territory; therefore, it is difficult to reconstruct and observe the tangible and intangible values of the overall development of local systems. In this sense, a differentiation between project territory and context territory is introduced. The project territory is a temporary territory configured through the Olympic project and works to realise the mega-event. On the other hand, the context territory, whether or not it is affected by the Olympic event, follows its own rules and tasks of ordinary transformation.

Representing a material, perceptive and symbolic transformation of the territorial space through a mega-event can be seen as a production of values typical of contemporary society.

Therefore, through observing Olympic Village planning models, we will introduce common patterns in terms of urban typology to classify the different experiences.

The aim of the following book is to analyse the different spatial transformations of the Olympic event, leaving the focus on the event and concentrating on the various Olympic Villages that have developed over time and remain a heritage available to most host cities. Therefore, the functions and reuse of the Olympic Village in the post-Olympic period in different territorial contexts will be analysed. Furthermore, given the summer and winter events, different territorial contexts that can absorb or reject temporary situations by elaborating new cycles of territorialisation will be observed. Metropolises are subjects that are most likely to absorb temporary events such as universal expositions, the capital of culture, the Olympic Games and other events involving new cycles of territorialisation.

The following manuscript can be divided into four parts.

The first part addresses conceptual, theoretical and methodological issues related to the general and specific objectives of the monograph. The first chapter attempts to provide an overview of the historical development of the mega-event and how it can be used as a tool to promote the territory internationally. The second chapter will look at the phenomenon of Olympic urbanism, given its evolutionary stages over time. Subsequently, Olympic territorialisation will be introduced, and an overview of the elements of impact, legacy, affection and sustainability will be provided to define the holistic concept of Olympic legacy. The chapter will also explore the evolution of the territorial dimension in view of the location of the event within the host cities or regions. In the third chapter, however, after an overview of the history and evolution of Olympic villages, the chapter will look at the spatial and urban organisation. By introducing the spatial models of the two editions, the reader will be able to observe the different spatial models adopted by the host cities in relation to the territory.

The fourth chapter allows us to examine five case studies selected in consideration of their territorial relevance. The transformation of accommodations from Turin 2006 to Beijing 2022 will enable the reader to observe the main changes and quantitative differences with the help of the common elements proposed within the 2022 PhD study. Finally, the last chapter will allow for a discussion on the evolution of Olympic housing between the twentieth and twenty-first centuries in consideration of the results obtained within the study. Furthermore, the following chapter will provide some indications regarding the future implications for implementing or reconsidering Olympic accommodation. The following text, with the results obtained within the doctoral study, will offer an innovative vision for the research and implementation of Olympic Villages.

CHAPTER I

Sports Mega-Events and the Promotion of the Territory

ABSTRACT

This chapter presents considerations on mega-events from a global development perspective. After illustrating the different types of events with contributions by Roche (2000), Guala (2002) and Getz (1997), we will look at the other relationships that Olympic events have developed in different parts of the world. The chapter will analyse Olympic cities according to location, population and number of participating athletes, reflecting on the globalisation of the event. The consideration of the mega-event as a catalyst for the transformation of places, through the contribution of (Kearns & Philo, 1993), allows for the study of the promotion of places from the perspective of the phenomenon of place selling. Finally, the chapter continues with the contribution of Kotler (1993), who analyses the concept of strategic management of the image of places from a global idea of strategic management of the image of places.

1.1. Mega sporting events from a global perspective

1.1.1. The different types of events

The definition of mega-event is a specific notion introduced since the first edition of the London International Exhibition in 1851. Meanwhile, the Greek society created the first concept of an Olympic event, which held the first edition of the Olympic Games in 776 BC. Another event that marks the history of events is the Christian Jubilee, held for the first time in 1300 and is still organised every fifty years. The Ancient Greek editions of the Olympic Games were indeed different events, held in other city-states, implying different religious and symbolic meanings of antiquity. After fifteen centuries of interruption, in 1896 Baron Pierre de Coubertin re-established the modern Olympic Games through new rules guaranteeing global participation. Therefore, since 1869, the modern Olympic Games have been organised every four years in different countries and cities around the world. In addition, 1924 was the starting signal for the first winter edition, which since 1992 will be organised two years after the summer edition in a different city and country.

Modern cities have harnessed and used mega-events to bring their social and economic role to different levels and scales. For example, the international exhibitions were first held in London in 1851[1], aimed to demonstrate and promote technological progress among the more developed nations through a political role. For the first time, train and boat trips were organised to see the scientific and technical progress of the English. Until the second edition of the Olympic Games in Paris in 1900, cities and local public administrations that wanted to organise a World's Fair promoted the image of cities open to change.

By analysing these historical assumptions, mega-events can be seen in a different context that will allow the discipline of territorial marketing to advance. "Events are scheduled or unplanned events that have a limited duration and are created for a specific purpose" (della Sala, 2022).

One of the characteristics of mega-events is the programming and the purposes that each administration wants to acquire for the territory. In phase, the Olympic, even the Olympic event, as in other global events, is subject to different pressures that can reach different meanings to the Olympic project, diminishing its substantial and long-lasting potential.

Moreover, the event loses its ephemeral characteristics when the works require an undeferrable timeframe for the organisation and the predisposition of the Olympic works on the territory. Over time, the Olympic works have almost always been permanent, especially the facilities and the Olympic Villages. Therefore, the permanent occupation of space makes it possible to implement a territorial project, unlike other ordinary projects, which remain unrealisable for various reasons. However, the mega-event highlights the city's perspective and changes the priorities of the host cities.

Other authors, such as Getz (1997, 2004, 2008, 2016), have investigated events through territorial marketing strategies. In 1997, Getz identified the

---

1   The 1851 event is recognised as the world's first tourist event.

concept of dynamicity of events, that is, the ability to evolve and manifest itself in different places in the world. Getz's studies allow us to identify events in eight categories:

- Celebrations (festivals, carnivals, religious events, commemorations);
- Artistic events (concerts, other shows, exhibitions, awards ceremonies);
- Business/commercial events (trade fairs, markets, meetings, conferences, conventions);
- Sports competitions;
- Educational and scientific events (seminars, workshops, congresses);
- Recreational events (games, non-competitive sports, hobbies);
- Political/civil events (inaugurations, investitures, visits of authorities, parades);
- Private events.

The classification of events was revised by Ferrari in 2002, including different variables that affect the preparation and management of the event, as well as the territorial consequences that the event itself may generate.

So far, the event, as defined until 2000, maintains a certain conceptual distance from the space of production. According to Getz, the place of the event plays an almost secondary role, favouring only the event's evolution in time. In reality, mega-events, as Dansero suggests, "need a space to manifest themselves" (Dansero, 2002). Therefore, the event can be observed as a specific point in space-time.

Mega-events in contemporary society often represent the gathering and overexposure of different values that follow different logics in space-time. The city is that which provides the space, the central government is responsible for the financing of the event, the Olympic world is accountable for the organisation of the event, the local community is responsible for the acceptance of the event and the whole world is that which defines the size and magnitude of the event on a global level.

Table 1. Classification of events (Source: Ferrari, 2002, p. 42)

Criterion	Ranking
Cadence	Periodic events; one-off events
Duration and timetable	One day; one month; type of day/hours
Area of attraction	Local; regional; national; international; global
Number of visitors	....
Type of access	Free; Paid; Partly paid
Level of media attention	Local; regional; national; international
Target audience	Young people, seniors, singles, families with children, etc. Business tourism, cultural tourism, etc. Residents; tourists Experts; non-experts Unique visitors; repeat visitors
Spatial areas of location	One zone, several zones, zones for special events, a city district
Location	One; multiple (travelling events)
Package of attractions offered	One event: one main event and a set of smaller events and other attractions; several smaller events
Services offered	Information; reservations; transportation; reception; catering; security; health; other events for entertainment, socialising or other purposes
Main objective	Increasing tourism flows; fundraising; entertainment; trade promotion; improving the locality's image; encouraging local participation in an activity; philanthropic; social; other.
Theme	....
Initiative and ownership	Promoters, creators, owners of the contents of the label
Type of organisation	Volunteers; professionals; public bodies; multiple parties; sponsors; suppliers; other
Main sources of financial resources	Audience; sponsorship; ticket sales; other
Sector	Fair; festival; congress; concert; sports competition; exhibition; meeting; religious festival; expo; commemoration; other; other
Typology	Celebrative; cultural; recreational; folkloric; labour; religious; sporting; commercial; political; fundraising; other

Table 2. Typology of mega-events (Source: Roche, 2002, p. 4)

Type of event	Example	Target/market	Media interest
Mega-event	Expo, Olympics, World Cup, Football World Cup	Global	Worldvision
Special event	Formula 1, regional sporting events (Pan American Games, European Cup)	World, Regional, National	National and international media
Hallmarks event	National sporting events (Australian Games), significant city sports festivals/events	National, Regional	National media
Community event	Local events	Regional/Local Local	Local media Newspapers

1.1.2. The mega-events

The definition of mega-events refers to events of short duration, unique and with a high media impact. Mega-events "enhance the awareness, attractiveness and profitability of a tourist destination in the short and/or long term" (Ritchie, 1990, 1991).

According to the contributions of Burns and Mules (1986), mega-events are considered special events: exceptional events capable of generating significant demand for services limited to a relatively short period that can attract and acquire new international funds for the development of the entire host region.

Over time, mega-events have developed into a global category within which it is possible to distinguish different aspects of their media implications. Roche (2000) advanced the classification of mega-events in consideration of the magnitude of the target market and media interest (Table 1).

In addition, to look at mega-events in-depth, we consider other elements provided by Guala (2002) about the magnitude and involvement of participants and the budget for the event organisation. The table provides an updated classification in 2015 by Guala (Table 2). Thanks to this double

Table 3. Objectives and budgets of some events (budgets defined in
2005) (Source: Guala, 2015, p. 45)

Event	Participation	Initial budget (€)
Summer Olympic Games	More than 10,000 athletes Olympic family 50,000 members 7,000-8,000 journalists	3,000-4,000 million
Olympic Winter Games	2,000-2500 athletes Olympic family 30,000 members 5000 journalists	800-900 million
G7/G8	Delegations, 5000 journalists	80-100 million
European City of Culture	Tourists, journalists	100-200 million
Tall Ships race	Visitors 1,000,000	500,000 thousand euros
Euroflora (Genoa)	Visitors 800.000	has not been detected
Motor Show (Bologna)	Visitors 1,000,000	has not been detected
Exhibition of Turner and the Impressionists (Brescia)	Visitors 500.000	has not been detected
Boat show	Visitors 350,000	has not been detected

classification, we will see other elements that help us to define mega-events
in depth.

Among the divisions observed in Table 2, *mega-events* and *hallmark
events* are those with particularities and characteristics of uniqueness, prestige
and value (Hall, 1989; Ritchie, 1990; Roche, 1992), considering their funding
and public attendance.

Distinctive *hallmark events* can be defined as recurring events linked
to a particular place that takes advantage of and enjoys some specific attraction elements. The *hallmark events* attract a global audience and many

visitors around a particular theme[2]. Therefore, the organisation of *hallmark events* is complex because of the number of visitors and national and international media involvement.

On the other hand, mega-events, in terms of visitor numbers and global audience, attract a high level of interest and attention on a continental level (Short, 2008). The preparation and planning of mega-events, being so complex, requires years of preparations and a vast involvement of human and financial resources. In addition, mega-events, having a global impact, need candidate cities to be prepared to host an event of such magnitude. Therefore, the celebration of these events takes place in tourist cities, metropolises, and capitals that can withstand the complexity of the event's impact (Hiller, 1998).

Host cities can use and enjoy their connecting and attracting power to leverage different existing territorial elements. Therefore, mega-events must be organised through other general objectives to be targeted globally. The possibility of hosting a mega-event triggers various territorial transformations to rebuild the host territory for a new position at the worldwide level. Repairing host cities involves improving transport systems, accessibility services, structures and infrastructures. Therefore, mega-events can be an excellent opportunity for host cities to rebuild their fabric and spatial system. Moreover, they can be an intangible opportunity for a change in the philosophy of intervention and a unique opportunity to renew themselves and propose new perspectives for future development.

However, mega-events have suffered and continue to suffer crises about the specific historical phase. For example, the World's Fair had to transform itself to face a crisis that manifested itself in Seville and Hannover. On the other hand, after Montreal, the Olympic Games had to change their business model to avoid disappearing again. The 1984 Los Angeles Olympics can be considered a turning point in history, described as the first Olympics to be planned and organised with private funding, mainly based on new sponsorship income and the sale of television rights. These two events have

---

2   Examples: the Cannes Film Festival, the Oscar ceremony, the Sanremo Festival, the Turin Book Fair, the Perugia Jazz Festival, etc.

been, and continue to be, at odds with each other over time on common themes that allow the World's Fair and the Olympic Games to be linked.

Through the contribution of Roche (2000), it can be seen how the Universal Expositions served as an excellent inspiration for the creation and re-establishment of the Olympic Games by Baron de Coubertin. The first editions of the Olympic Games (Paris 1900, London 1908) took place in the same year as the World's Fair, including some rituals such as the opening and closing ceremonies. According to this interpretation, many twenty-first-century events are products of the Expo and the Olympic Games. Local events continue to draw inspiration from the mega-events to reproduce their philosophy and transmit intangible values to the local community.

1.1.3. The Olympic map

As we have seen in the preceding paragraph, mega-events such as the World's Fair and the Olympic Games have developed through totally different geopolitical dynamics. The Expo was born in the nineteenth century and was developed in a bourgeois and colonising context. The World's Fair was intended to demonstrate and communicate the economic power of states through trade and technological progress. The chosen cities became new urban symbols that encouraged intense competition between the world's capitals, to the point of creating a body to regulate the organisation of the event (*the Bureau International des Expositions*).

Therefore, the exposition can be defined as an event that boosts new economic flows in the host cities. Over time, the cities that have hosted Universal Expos have tried to enter into the global city networks outlined by Professor Sassen (1991). On the other hand, the modern Olympic Games were born in 1896 and developed in a bourgeois context, with Baron Pierre de Coubertin as the first sponsor of the revival of the Games. The Olympic Games were intended to raise awareness and communicate messages of fraternity, effort, loyalty and participation through sport. Over time, the chosen cities have become new urban symbols encouraging competition between states, to the point of boycotts by the world's major economic powers.

Therefore, the Olympic Games can be defined as a geopolitical event that boosts new economic flows and media exposure of the Olympic city worldwide. The network of Olympic cities is made up of "nodes" (the selected and excluded localities) and "networks" (the different relational flows that the nodes activate) (Dansero, 2002). The map representing the "nodes" generates an "Olympic map" where the large metropolises of the developed countries can be observed. "The countries that have hosted the Olympic Games belong to the usual G7" (Dansero, 2002).

In 1924, the Olympic Games were transformed into two completely different events: the summer and winter editions. Since 1992, the two

Table 4. Population and athletes of the editions of the Winter Olympic Games (Own implementation)

EDITION	POPULATION OF THE CITY	ATHLETES	% POB/ATL
OSLO 1952	447.000	694	0,16%
SQUAW VALLEY 1960	4.000	665	16,63%
INNSBRUCK 1964	100.000	1091	1,09%
GRENOBLE 1968	180.000	1158	0,64%
SAPPORO 1972	1.000.000	1006	0,10%
INNSBUCK 1976	117.000	1123	0,96%
LAKE PLACID 1980	5.000	1072	21,44%
SARAJEVO 1984	448.000	1272	0,28%
CALGARY 1988	640.000	1423	0,22%
ALBERTVILLE 1992	20.000	1801	9,01%
LILLEHAMMER 1994	23.000	1737	7,55%
NAGANO 1998	361.000	2176	0,60%
SALT LAKE 2002	174.348	2399	1,38%
TURIN 2006	900.000	2508	0,28%
VANCOUVER 2010	603.400	2566	0,43%
SOCHI 2014	364.000	2780	0,76%
PYEONGCHANG 1918	43.600	2833	6,50%
BEIJING 2022	19.638.000	2871	0,01%
MEDIA	1.392.686	1.732	3,78%
MAX	19.638.000	2.871	21,44%
MIN	4.000	665	0,01%

editions have become two various events organised yearly. As a result, the IOC has expanded its influence and magnitude worldwide. However, there are still some differences between the two types of events. A summer edition is an event for the world's major metropolises, while the winter edition requires cities with a minimum climate, temperature and altitude for the competitions.

Therefore, the Olympic map and nodes have been expanded only on a few occasions, when geo-economic reasons and the phenomena of geographical expansion of trade have prevailed (Tokyo 1964, 2020, Moscow 1980, Seoul 1988, Sydney 2000, Beijing 2008, 2022, Sochi 2014, Rio 2016, PyeongChang 2018,). Tables 4 and 5 show the editions of the Olympic Games held to date, the population of the city and the athletes participating in the summer and winter editions.

Figure 2 allows us to observe the evolution of the population in the summer towns, which, from 1936 up to 2020, has evolved by 17.57 per cent.

On the other hand, Figure 3 shows how the evolution of the population in winter towns from 1952 to 2018 has evolved by 1.33 per cent.

POPULATION TRENDS AT THE SUMMER OLYMPICS

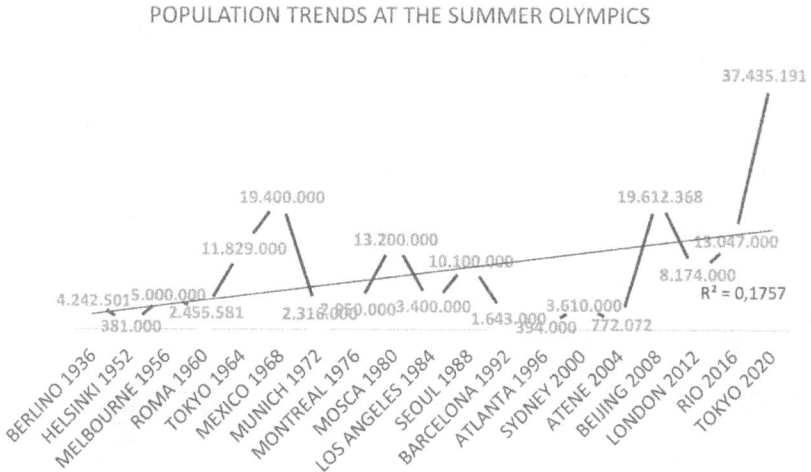

Figure 1. Evolution of the population in the Summer Olympics (Source: Own implementation)

Table 5. Population and athletes of the editions of the Summer Olympics
(Source: Own implementation)

EDITION	POPULATION OF THE CITY	ATHLETES	% POB/ATL
BERLIN 1936	4.242.501	3963	2,20%
HELSINKI 1952	381.000	4955	4,24%
MELBOURNE 1956	5.000.000	3155	0,19%
ROME 1960	2.455.581	5338	0,17%
TOKYO 1964	11.829.000	5151	0,03%
MEXICO 1968	19.400.000	5516	0,07%
MUNICH 1972	2.316.000	7134	0,31%
MONTREAL 1976	2.950.000	6084	0,21%
MOSCOW 1980	13.200.000	5179	0,04%
LOS ANGELES 1984	3.400.000	6829	0,20%
SEOUL 1988	10.100.000	8391	0,08%
BARCELONA 1992	1.643.000	9356	0,57%
ATLANTA 1996	394.000	10318	2,62%
SYDNEY 2000	3.610.000	10651	0,30%
ATHENS 2004	772.072	10625	1,38%
BEIJING 2008	19.612.368	10942	0,07%
LONDON 2012	8.174.000	10567	0,13%
RIO 2016	13.047.000	11238	0,09%
TOKYO 2020	37.435.191	11000	0,03%
MEDIA	8.419.038	7705	0,68%
MAX	37.435.191	11238	4,24%
MIN	381.000	3155	0,03%

Finally, (Table 7) shows the world geography of the Olympic Games and how this needs to respect the globality of the event and the philosophy of exporting the Olympic model to the five continents. For example, Africa has never been able to host an Olympic event. However, after London 2012, the Olympic event was moved eastwards, only to return to Europe

POPULATION TRENDS AT THE WINTER OLYMPICS

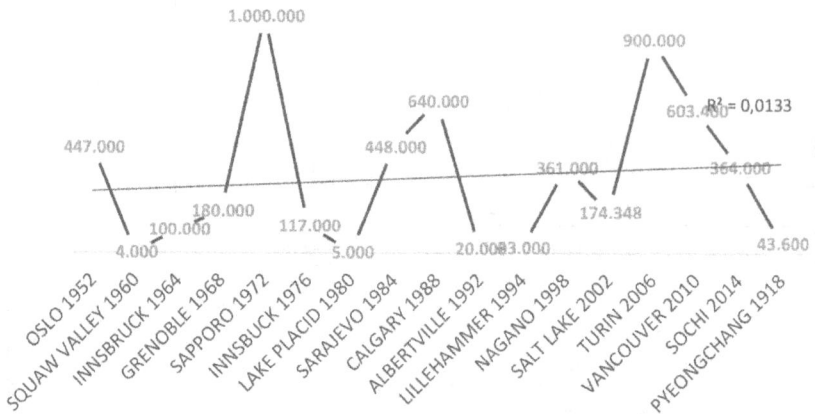

Figure 2. Evolution of the population at the Winter Olympic Games (Source: Own implementation)

Table 6. Organisation of the Olympic Games in the world (Source: Own implementation) (The table does not include the assigned editions of Paris 2024, Milan-Cortina 2026, Los Angeles 1928, Brisbane 2032)

Continent	Summer Olympics	Winter Olympics	Total
North America	5	6	11
Central and South America	2		2
Asia	5	5	10
Australia	2		2
Africa			0
Europe	16	13	29
TOTAL	30	24	54

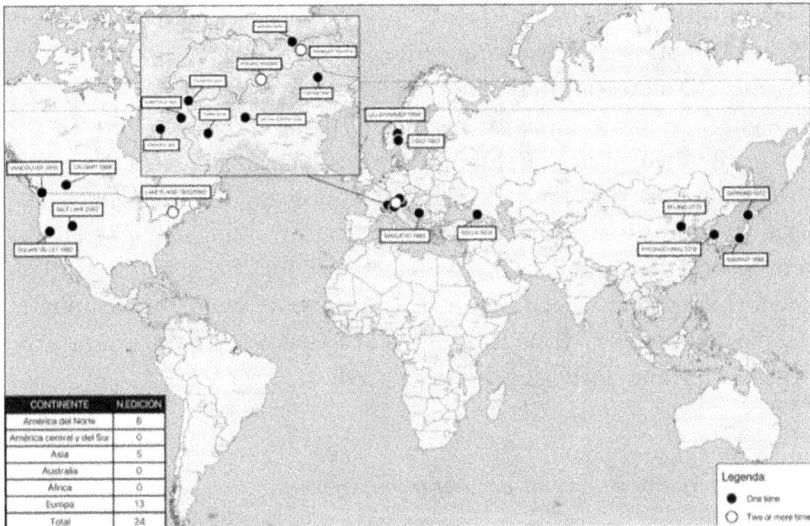

Figure 3. Map of the Summer Olympics (Source: Own implementation) (The map includes the assigned editions of Paris 2024, Los Angeles 1928, and Brisbane 2032)

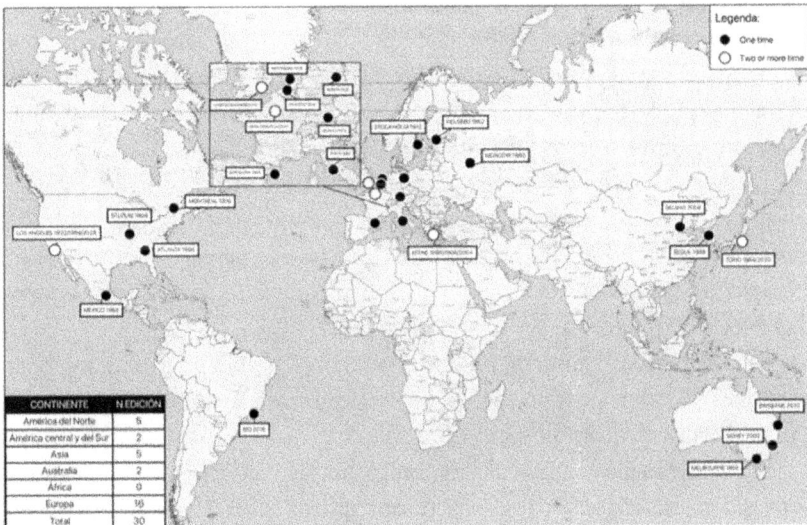

Figure 4. Map of the Winter Olympic Games (Source: Own implementation) (The map includes the assigned editions Milan-Cortina 2026)

in 2024 with the allocation of Paris. Thus, for the first time in the history of the Olympic Games, five consecutive Olympic Games were held outside Europe. The increasing number of disciplines, new sports and athletes has brought new uncertainties for the future Olympic events in Europe. The choice of Olympic cities in Europe follows the patterns of the consumer society, supported by the main Olympic sponsors. Moreover, the mediatisation of the event and the growth of private funding have profoundly transformed the Olympic model worldwide, altering the intangible symbols and meanings at the core of Baron de Coubertin's Olympic philosophy. In the next section, we will analyse the role of local communities during the organisation and hosting of the mega-event.

1.1.4. The role of local communities in mega-events

As outlined above, the Olympic Games can be seen as an exogenous phenomenon that takes advantage of the territory to manifest itself and develop new processes of local territorialisation. The mega-event is a complex mechanism that involves the involvement of global elites and falls disproportionately on local elites. The general perception is that the mega-event turns a chosen place into a "commonplace" (Dansero, 2002, 2014) for the use and consumption of sponsors, the International Olympic Committee and the media.

However, each edition has its particularities and specificity. In addition, the event space keeps changing and expanding over the editions. The IOC tries to standardise the processes and phenomena of the event, but each venue has its own identity and morphology, making them incomparable over time.

The Olympic event can be understood as a global phenomenon that provokes different productions in each specific place, incorporating itself into a worldwide homologation order.

Local strategies must find a meeting point for reducing land consumption and producing new values and opportunities for the target territory. As the venue of the event, the regional context should be considered the leading actor for the possible relapse of the Olympic effects in the post-Olympic period. Analysing the territory and focusing on shared objectives seems to be the only accessible option for implementing local development processes that can rebuild and improve our cities through national pride. The Olympic Games, like the World's Fair, seek to promote developments in the field of technology through sports facilities, equipment, structures and infrastructures. The choice of the sites chosen to host the Olympic event must be widely shared by the local actors. In this context, the possibility of proposing surveys and conferences is introduced to encourage the information and involvement of citizens in the bidding process. The following activities presuppose the existence and validity of a local elite adequately equipped both to filter the event's homologating tendency and interpret the need for change in the territory in the long term (Scamuzzi, 2002).

Local actors must consider and recognise the territory as tangible and intangible resources that can be exploited thanks to the mega-event. Furthermore, in the post-Olympic period, the territory will have to integrate and relate to global networks.

Spatial transformations on a global scale require additional financial resources to be provided by the central government in a project of new territorial connections and interconnections. The world's cities and significant metropolises can only implement works and infrastructures with strong and shared support from the central government.

In addition, the event planning must consider real estate interests and speculation on the sale of the land chosen for the construction of the Olympic venues[3].

Roche (2000) summarises some of the instrumental uses of the Olympic event for the world's elite:

- The mega-event can be interpreted as a "theatre of power" for the promotion of dominant ideologies,
- The mega-event can be used to define a change of strategy,
- The mega-event can serve as a continuation of a past heritage.

---

3   The analysis of these dynamics can be deepened through the *growth machine* model.

Therefore, the mega-event can be used to reconstruct and redefine the territory.

The following instrumental use has allowed cities such as Rome, Tokyo, Barcelona and Turin to redefine their territory through the predisposition of Olympic works. Today, this model is not used due to its enormous complexity and the effort involved by central governments. Moreover, the growth of the "no Olympic" movements has made the bidding process and the promotion of the major metropolises more complex.

The case of Barcelona 1992 is emblematic from the point of view of urban regeneration and the global repositioning of the city as a brand. The Barcelona project carried out some ambitious urban operations to open its doors and coastline to the world. A change in the urban hierarchy has definitively transformed the territory and space of the Catalan capital. In addition, a series of events, such as the Forum of Culture in 2004, has made it possible to continue with a progressive urban renewal that continues to this day. However, the exploitation of the city at a global level has meant greater security for local actors in securing and acquiring new international events. Including the Olympic city in a global network can be an extraordinary opportunity for the town and its citizens. The definition and planning of a more comprehensive project can allow for a general improvement of services and facilities while respecting the real needs of the citizens (della Sala, 2022).

The event's timing leads to the stratification and centralisation of endogenous forces around a single objective for ten to fifteen years: to prepare all the works for the Olympic event. The centralisation of national forces to comply with the IOC rules will slow down the ordinary progress of the Olympic Games and will dispense a lot of resources for the execution of the Olympic works.

Mega-events indeed destabilise the day-to-day management of the territory and the execution of infrastructure works. Moreover, the management of the event requires different plans for the organisation of the Olympic event:

- Application process,
- Pre-event planning,

- Organisation and execution of the event,
- Post-event planning and organisation.

In the post-event planning and organisation phase, the permanent effects of the spatial and social transformations induced by the event on the local community will be observed.

1.1.5. Mega-events and the transformation of places

Organising a mega-event creates new expectations and concerns in the local territory. The Olympic event should be observed through two different visions: a regional vision, as a valuable opportunity to establish new local synergies and alignment of very ambitious territorial projects without Olympic funding, and a global vision, as an opportunity to create new infrastructures and sports facilities to achieve new opportunities about development and the improvement of citizens' well-being. In addition, the attraction of extraordinary financial resources can serve as a driving force for a new image of the city on a global stage, which would otherwise be impossible (della Sala, 2022).

This double vision can be interpreted as a project scale representing some critical nodes mainly linked to the territory and significant territorial transformations fundamental to hosting the mega-event. Moreover, over the years, the issue of environmental impact has been established as the third essential element for developing the Olympic territories. The potential adverse economic, social and image-related repercussions observed in Montreal, Sochi, and Rio have led to these cities receiving much criticism from the international community and continue to be the subject of debate[4]. However, the overemphasis on the event can lead to increased costs and overestimation of the capacity of the venues. In addition, the lack of consideration of local priorities and social requirements has profoundly transformed the city selection process. For the first time in its history, the

---

4   In recent years, the IOC has received a lot of criticism regarding the costs and uses of Olympic facilities in the post-Olympic period.

IOC is now selecting cities that propose an average percentage of 90 per cent of temporary or existing venues[5]. This new development model has changed the transformations of Olympic venues through the mega-event. The preference for cities already equipped or with possibilities to take advantage of temporary technology will be the dominant strategy for the next twenty years regarding the future of the Olympic event. Thus, the tremendous territorial transformations of the twentieth century were related to industries. In contrast, the territorial transformations of the twenty-first century have become a phenomenon closely linked to mega-events as they are considered an acceleration process for the host site (della Sala, 2022).

Like the Expo, the Olympic Games contribute significantly to leaving symbols, monuments and heritage in the places hosting the event. The Palais de Cristal and the Eiffel Tower for the World's Fair, the ski jumping slope in Oslo, the Stadium in Beijing, and the Olympic Village in Munich are symbols of the urban transformations linked to the mega-events. However, sometimes symbols can be temporary, as with the *Medal Plaza*[6]. In Turin, it remains a fundamental piece in the memory of the citizens.

Therefore, the organisers' objectives may differ, but urban change due to mega-events is undoubtedly an effect that was observed from the beginning and has manifested itself through different types of urban impacts on the host territory.

The contribution of the geographers Essex and Chalkey (Essex & Chalkley, 1998) allows us to observe the different transformation processes that have emerged over the years and in their various phases. Phases 7 and 8 were added to update the geographers' contribution to new events observed over the years.

1. *Zero impact→* The first editions of the Olympic Games, organised in a single sports facility, did not generate significant urban transformations. Existing sports facilities were used.

5   The new IOC rules were published in 2018, and the first city chosen with the next update was Brisbane 2032.
6   For the first time in Olympic history, Turin's Olympic Square was planned outside the Olympic Village.

2. *At the turn of the century*→ The Olympic Games of this period provided new sports facilities with limited impact, and for the first time, in Los Angeles, the first accommodation structure for Olympic athletes was built.

3. *Olympic gigantism*→ After the Second World War, the impact of the Olympic Games started to become more significant in terms of the number of sports facilities and Olympic districts with new transport systems attached. The increase in urban transformations achieved different results, and the first failures, such as Melbourne and Montreal, began to be observed. On the other hand, cities such as Oslo, Rome, Tokyo, Mexico City and Munich took advantage of the facilities and infrastructure developed for the mega-event as an accelerator of processes included in their development plans.

4. *The way out of the Olympic crisis*→ The increase in the size of the Olympic works, and the failure of Montreal opened a period of crisis that ended only with the success of the Los Angeles edition in 1984, an edition that will be remembered in history as the first to be developed through temporary financing and which used university facilities and accommodation. At that historic moment, the IOC introduced the top sponsor programme to increase funding for the Olympic movement and not allow the event to disappear as it had done in the past. The success of the Los Angeles event opened the door to the commercialisation and globalisation of the Olympics.

5. *Expansion to new markets*→ The Seoul edition sought to accelerate cultural and economic processes through profound transformations to meet Western standards. Seoul (1988) took advantage of the Olympic opportunities to adapt and improve its urban infrastructure.

6. *Urban metamorphosis*→ The 1990s were inaugurated with the Barcelona Olympic Games, which remain a model and an example of using mega-events as a catalyst for urban transformation and renewal. The experiences of Sydney and Turin follow

the same principles of the Barcelona model with the addition of elements of environmental impact and sustainability.

7. *The Olympic Village as a central element of Olympic urban planning*→ The editions of Turin, Beijing, London, Sochi and Rio have provided an opportunity to reflect on Olympic accommodation and the issue of housing in the twenty-first century. These years are represented by different housing projects with mixed functions through the support of new sporting and sustainable neighbourhoods. At the same time, they have allowed for a debate on the issue of housing and the construction of new sports neighbourhoods.

8. *The temporary event*→ New criticism regarding the size of the Olympic project and housing has caused a new crisis for the IOC. The development of associations against the Olympic event and the decrease of free space in the candidate cities has led to an inflexion in the demand for bidding to host the Olympiad. In addition, the infrastructural costs observed in Sochi and Rio have been criticised for their uselessness in the post-Olympic period. The disuse and neglect of sports facilities and the Olympic Village that emerged in Rio undoubtedly changed the event's focus over time. In 2017[7] for the first time in Olympic history, the IOC allocated the Summer Olympics directly to cities that have hosted the Olympics in the past (Paris and Los Angeles), taking advantage of the reuse of existing sports facilities and relying on temporary structures. This organisational change has introduced new elements in the allocation phase of the Olympics that continue transforming the event into something ephemeral for the citizens. Therefore, the mega-event will take advantage of the urban space in a temporary way without leaving a material legacy in the city, as was observed in winning editions such as Oslo, Rome, Tokyo, Barcelona, Sydney, Turin and London.

---

7   The Lima session in 2017 was pivotal for the organisation and planning of future Olympic editions.

After observing the different transformation processes that have emerged over the years, in the following section, we will look at the mega-events from the perspective of a tool for promoting territories and Olympic sites.

1.1.6. Mega-events from the perspective of territorial promotion

Territorial promotion is a phenomenon that involves and has extensions in different economic and social activities carried out by individuals and organisations that manage places (Kearns & Philo, 1993). Places can be urban, peripheral, rural or regional areas.

Promoting places is a phenomenon that first appeared in the Western world and then developed at the end of the twentieth century in many cities in underdeveloped and Eastern countries. Territorial promotion makes it possible to adopt different strategies to promote other development models in places. Size, economics, history, habits and morphology are some factors to be considered in promoting the territory.

The concept of "territorial marketing" in recent years has produced some references that we will take as a basis for our analysis: Logan and Molotch, *Urban Fortunes* (1987); Kotler, Marketing *place* (1993); Kear and Philo, *Selling place* (1993); Ward, *Selling place* (1998).

Promoting places involves different forms of organisation and participation of public and private subjects. Public and private bodies strive to sell the image of a specific geographically defined place, usually a city, to make it attractive to economic enterprises, tourists and even the inhabitants of that place (Harvey, 1991). One of the main objectives of territorial promotion is to stimulate companies to establish themselves in that place and to promote it in terms of tourism so that constant capital investment can be ensured to create jobs and new activation of the local economy. Thus, since Barcelona 1992, the Olympic Games have been transformed by an essentially economic logic for selling the sites as a global brand. The selling of places involves conscious and deliberate manipulation of culture to increase the attractiveness and interest of places (Kearns & Philo, 1993).

On the other hand, the promotion and manipulation of places depend on promoting traditions, habits, lifestyles, architecture, art and, supposedly, the places themselves. In this sense, the selling of places is manipulated due to the implication of a series of values and objectives related to events, exhibitions and other cultural activities that have no necessary association with the specific place and could, therefore, be considered inauthentic and ephemeral (Larson, 2009). Considering the importance of the history of places in citizens' self-identification, it is inevitable to consider the specific heritage to manipulate the event about the context territory.

Therefore, promoting places focuses on culture, history and capital and how these elements are mixed in selling places (Kearns & Philo, 1993).

1.1.7. The promotion of places

Promoting places in the twenty-first century has become a fundamental element in promoting contemporary urban societies. This can be seen as a social and economic opportunity in the wild jungle of the capitalist market and consumer society. Thus, places become products or commodities. History, culture and economics make some places more attractive than others and can be marketed about the demand and supply of global tourism (Montanari, 2002). The promotion of places is embedded in the host places' physical and cultural transformation. Most post-industrial cities have become products to be sold in different markets interested in that specific place. There is a direct sense of selling the postmodern city, which involves deliberately creating cultural-historical packages. More or less obvious combination of cultural and historical elements to produce marketable products, such as theme parks (Sorkin, 1992), the simulation of past rituals or events. The indirect promotion of place involves subtle promotion with cultural and historical materials in producing what are supposed to be attractive, pleasant and uplifting environments (Kearns & Philo, 1993). From this starting point, the modern city is an excessive decontextualisation of places' cultural, historical and social heritage in a local context.

In the twenty-first century, it is crucial to consider all types of memory associated with the different local contexts that recall the history and culture of each person. The manipulation of the culture and history of places will undoubtedly bring economic benefit to some of those concerned. Including local peculiarities and specificities in a planning strategy favours territorial promotion without undermining the history and culture of places.

1.1.8. Reconstructing the image of industrial cities

Since the 1980s, place promotion, first in the United States and then in Europe, has become a core activity in place redevelopment, making the redevelopment project more complex and involving more professionalised resources to compete with other places. Today, land marketing in the United States is a multi-million-dollar industry, as many consultants and public relations firms specialise in tourism, advertising and selling cities, states, retirement communities and resorts (Holcomb, 1990). The main objective of place marketing is to build a strategy to develop a new image of places and to replace the old negative images left by *Fordism*.

Industrial cities mainly use this tool to build a new image that can replace the old negative photos, seeking to create new ephemeral images preferred by the market. Industrial cities in the twenty-first century have restructured themselves to develop a new service economy in the post-industrial era.

Glasgow is recognised as the first major post-industrial project that allowed the city to build a new image within the world of events (Garcia, 2005). Barcelona, Turin, Milan and London are some Olympic cities that have used mega-events to build a new global post-industrial image of the city. The packaged image of the promoted venues reflects the aesthetic tastes of postmodern society. As Harvey anticipated, aesthetics has triumphed over ethics as the main focus of social and intellectual concerns (Harvey, 1991).

For many authors such as Sennett (1987), Sorkin (1992), Sassen (2018), and Davis (2015), the city of Los Angeles is the paradigm of the myth of a city in constant change that has helped to fuel the material transformations of the landscape, the image of the town and has shaped its metropolitan transformations. Los Angeles is an excellent example of how real estate, cultural and economic speculation is inevitably embedded in large-scale urban development. Therefore, economic, cultural and political processes have been organised through different strategies for promoting and commercialising urban space.

"A given built environment expresses specific patterns of production and reproduction, consumption and circulation, and as these patterns change, so does the geographical pattern of the built environment" (Smith, 1996). Therefore, any city's urban and social space is constantly changing, introducing new forms and structures that abandon the historical forms of places (Gehl, 2006). Both physical and metaphysical processes of transformation are constantly evolving, producing further modifications to a particular space for a new system of production and consumption of places. *Canary Wharf* in London is one of the first projects in the 1980s driven by the reconstruction of financial services through private funding and public subsidies. Today, *Canary Wharf* has regained its central role in the metropolitan city, promoting the project as an example of post-industrial development, privileging private objectives and property speculation. In such a way, the architecture of places has once again become a new form of advertising the processes of redevelopment and promotion of places (Muñoz, 2015). In recent years, *archistar* mega-projects have become catalysts for other processes.

While the advertising of places indicates a specific geographical context, architectural forms are permanently inscribed in a precise place, feeding the desires of consumers.

Venturi (1966) advised architects to concentrate on decorating the surface of buildings, turning the environment into a massive advertising space displaying images and symbols selected by the architects. According to Venturi, architecture should be seen as the essence of symbolic manipulation rather than the production of forms.

Thus, in the modern era, architecture is seen as an effective instrument for social, cultural and economic change that satisfies citizens visually (Charlesworth, 2005).

1.1.9. Towards an ephemeral city

The ephemeral city is identified in the lack of a global vision and col-
laborative strategies, characterising the postmodern city through a proliferation of different nodes and symbols (Soja, 2000). Cities have been
transformed into a set of closed architectures that are designed as autonomous entities, each projecting different values and symbols within
the city[8]. The consummated space produces a selectivity of functions and
social uses. In such a way, areas are transformed into places of architectural spectacle where aesthetics are more important than function and
integration into urban planning (Zukin, 1991). Modern cities become
imaginary cities that are visually stimulating and reflect postmodern aesthetics by their power to re-urbanise entire areas, regardless of the landscape context.

1.1.10. Mega-events and city promotion

As we have observed in the preceding chapters, entertainment, culture,
and services are critical components for successfully reconstructing the
image of post-industrial cities. The attraction of tourists and the free promotion of athletes and members involved[9]. The organisation of mega-
events have become a central element of the new urban economy founded
on services. In addition, mega-events can promote a change in the perception of places through global audience and media coverage. However,
not all cities can profit from the promotion of venues. In some cases, the
place promotion strategy can enhance national perceptions of the specific place, favouring other benefits related to promotion. Thus, since the
twenty-first century, cities have started to bid for mega-events to promote
venues through Olympic bids. Promoting Olympic towns that will not

8   Muñoz (2008), in his concept of "Urbanalización", advances some patterns for
     interpreting urban plans, describing them as non-cities, banal cities deprived of
     specific identity, submerged in individualism, based on speculation.
9   Sydney, in 2000, to promote the Olympic sites, organised two weeks of free trips
     for the top 500 managers in the world, offering different activities and visits.

ultimately host the Olympic Games has become another possibility for cities to rebuild and promote a new image of places through the bid alone (Muñoz, 2011). Virtually all contemporary metropolises have essential cultural attractions, such as museums, galleries, theatres, facilities, and sand stadiums, which are the legacy of the bidding process or the hosted events (della Sala, 2022).

Regarding the cultural promotion of places, the organisation of sporting events in large contemporary metropolises has enjoyed a much higher level of interest than traditional cultural promotion. However, sporting events with a popular dimension make it easier for administrations to justify interventions at the local level (Smith, 2012).

As has been observed in many cities, business investments and sports programmes can enhance the city's image in the global competition of venues. In addition, the global size of sporting events helps administrations and businesses reinforce their commitment to the venues through critical interventions to host the event. The event's media coverage and broad marketing ensure that the city's name receives broad media exposure that could not be achieved through other types of events. For example, in the United States, cities and sports teams are strongly linked to promoting symbolic places transformed by the franchise. Thus, teams were built where the name of the place was only part of the team's identification, but without losing the team's identity.

On the other hand, in Europe, the teams have been incorporated into the tourist promotion of the city in a generalised image of the places. Despite these significant differences between countries, the power of attraction of mega-events is very similar worldwide. Therefore, even if nominations rise and fall depending on the allocation process, sports and mega-events will maintain or increase their importance in selling and promoting the post-industrial city. As for the contents concerning the city's promotion, we can only hope and wish for conscious exploitation by the organisers of the different dimensions of the city and the territories. The essence of reinvention consists of a process of change in existing perceptions through other values and symbols that will be instilled in the collective image of places (della Sala, 2022).

1.1.11. Local pride

One of the most essential aspects of place promotion is local approval and enthusiasm in new host cities[10]. The concern about providing global projects that respect local scales is one of the most exciting issues about territorial marketing in recent years. Residents in the city and nearby local communities are those consumers who can ensure long-term exploitation and positive results (Chen, 2013). Organising a mega sporting event such as the Olympic Games bases its engagement with local communities on proposing a widely accepted bid that engages with local associations to create joint strategies for the event's organisation. However, over time, we have observed different projects and results that sought local complicity and effort to commit to the post-industrial revaluation of the territories. The promotion of the new image of the post-industrial city must be approved by the general population to create new strategies that can be implemented in the post-event period. However, the new territorial marketing model is in total contrast to the industrial model. The post-industrial city model observed in the United States is still seductive for local organisers. "Post-industrial city marketing is essentially an American invention" (Ward, 1998).

Territorial marketing can only target specific social groups with an ethnic targeting approach. This ultra-segmentation strategy is leading to new changes in the organisation of mega sporting events.

1.1.12. Strategic management of the image of places

Strategic image management (SIM) is a continuous process of researching the image of a specific location by analysing its audience. Segmenting and targeting your particular image through positioning an existing image or

---

10   For example, the participation of citizens in volunteering is a fundamental element for the acceptance of the event in the host community. In Barcelona, there were 40,000 volunteers; never before 1992 had there been such a high level of citizen participation and involvement in a mega sporting event.

supporting the creation of a new image will communicate results to target audiences (Kotler, 1993).

How can the image of a place be measured? Kotler (1993) advises selecting a specific audience with some common characteristics.

Therefore, the first stage consists of audience segmentation and targeting:

(a) Resident,
(b) Visitors,
(c) Factories,
(d) Headquarters and corporate offices,
(e) Entrepreneurs,
(f) Investors,
(g) Foreign buyers.

Audience segmentation is most useful when it has six characteristics:

(a) Mutual exclusion,
(b) Comprehensive,
(c) Measurable,
(d) Accessible,
(e) Substantial,
(f) Differential response.

Subsequently, audience measurement should be developed by semantic differential:

(a) Develop a set of relevant dimensions,
(b) Reduce the set of suitable dimensions,
(c) Administer the instrument to a sample of respondents,
(d) Averaging the results,
(e) Checking the variance of the image.

Meanwhile, for an image to be effective, it should meet the following criteria:

- Validity,
- Credibility,
- Simplicity,
- Attractiveness,
- Distinctive.

Therefore, the tools to communicate the image will be provided through slogans, themes, symbols, events and campaigns specific to the segmentation of your audience[11].

In such a way, the distribution of the image of places should provide the target audience with the tools to influence the public in choosing the city as a preferred place[12].

Kotler (1993) suggests some tools for direct communication of the promotional message:

1. Effectiveness of guidance,
2. Personalisation of the message,
3. Interactive quality,
4. Measuring the response,
5. Relationship building.

According to Kotler (1993), tourism market segmentation should be provided through the following elements:

- Attractions sought: Sport, sun, leisure, play, culture,
- Market area/location: foreign, national, regional, local,
- Client characteristics: age, income, single, professional, religious,
- Benefits: price, quality, food, service, facilities.

---

11   At the Olympic Games, the message that the host city wants to promote to the world will be crucial in exploiting a new image. Beijing's 2008 message, "One world, one family", was one of Olympic history's most influential media slogans.

12   Meanwhile, eliminating a negative image is a long process of participation where all the main actors can recognise themselves in positive elements, eliminating the negative.

The choice of a target audience, the segmentation of the message, effective communication and the support of the tourism market can be considered elements that will help cities to define new strategies position and effectively communicate their places in the global tourism market (Hall, 1992).

According to Kotler (1993), there are some fundamental elements for the improvement of the competitive position of cities:

Basic information for strategic companies:

    I. The local labour market,
    II. Access to customer and supplier markets,
    III. Availability of sites, facilities and infrastructure for development,
    IV. Transport,
    V. Education and training opportunities,
    VI. Quality of life,
    VII. Business climate,
    VIII. Access to R&D facilities,
    IX. Availability of capital,
    X. Taxes and regulations.

Characteristics of the location:

    I. Labour tax climate,
    II. Incentives,
    III. Services,
    IV. Higher education, schools, regulation,
    V. Energy,
    VI. Communication,
    VII. Business.

On the other hand, places should not only attract tourists, businesses and investors but should also be committed to developing policies and plans for residents as part of a comprehensive and viable community acceptance strategy.

The key challenges:

    I. Places are increasingly threatened by the rapid pace of change in the global economic, political and technological environment.

    II. Places are increasingly at risk due to the normal process of urban evolution and decay.

    III. Places face growing competitors in their efforts to attract scarce resources.

    IV. Places increasingly have to rely on local resources to cope with growing competition.

    V. Places need to establish a strategic vision to meet these challenges.

    VI. Places need to establish a market-oriented strategic planning process to meet these challenges.

    VII. The site must adopt an actual market perspective toward its products and customers.

    VIII. The site must build quality into its programmes and services to complement another site.

    IX. Places must be able to effectively communicate and promote their competitive advantages.

    X. Places must diversify their economic base and develop mechanisms to adapt flexibly to changing conditions.

    XI. Places should develop and foster entrepreneurial characteristics.

    XII. The place must rely more on the private sector to carry out its tasks.

    XIII. Each location must develop its change processes due to differences in local culture, politics and leadership processes.

    XIV. The site must develop an organisational and procedural mechanism to sustain its development and maintain momentum once initiated (Kotler, 1993).

In conclusion, strategic place management is a fundamental tool for conveying and transferring host communities' message(s). However, extensive promotional practice can undermine community acceptance by introducing new criticalities in the context of participation and support.

CHAPTER 2

Olympic Urbanism

ABSTRACT

Having introduced the key concepts, the next chapter will allow us to examine the relationship between the Olympic event and Olympic cities. The next chapter will let us explore the relationship between urban transformations and Olympic cities, based on the contributions of Essex and Chalkley (1998), who consider the urban impact of the event in the context of physical transformations of Olympic cities. Meanwhile, the contribution of Kassen-Noor (2013) established concepts related to the transformation of the transport system in candidate cities in the context of spatial development.

2.1. Olympic cities

2.1.1. Introduction to Olympic cities

The Beijing edition in August 2008 marked a change in the global audience for mega-events. The Chinese event set record revenues from selling and marketing audiovisual rights[1]. Therefore, since 2008, with the advent of new digital platforms, revenues and viewership continued to evolve and contribute significantly to the global event audience. According to the IOC marketing report published in 2021, London 2012 and Rio 2016 had an audience of almost five billion people worldwide. In parallel, the Sochi 2014 and PyeongChang 2018 Winter Olympics reached an audience of 2.5 billion total viewers[2]. The following data allow us to observe a great distinction between the summer and winter editions[3]. For this reason, the

---

1   Sands, L. M. (July–August 2008). The 2008 Olympics' impact on China. *The China Business Review*.

2   CIO, Marketing Fact, 2021, pp. 8—21.

3   In addition, there is a 100 per cent difference in the marketing of audiovisual rights between the two editions.

difference between audience and funding from the IOC and Olympic sponsors allows us to assume that the winter edition is at a disadvantage. Therefore, the Organising Committees must provide complex projects with weak economic flexibility. Consequently, the dependency relationship between the city and the Olympic Games only continues with the summer edition. As we will observe in the next chapter, the winter edition, starting in 2006, has profoundly changed the size and exploitation model of the Olympic event. Earlier, in October 2009, two economists, Mark Spiegel of the Federal Reserve Bank of San Francisco and Andrew Rose of the University of Berkley, conducted a study on the Olympic legacy of host cities, focusing on strategic planning and highlighting its centrality in all aspects of the event's success[4]. Therefore, given the significant infrastructure investment, it is crucial to emphasise the importance of preparing a strategic plan in collaboration with all regional, national and international stakeholders.

Furthermore, strategic planning[5]. This allows for a legacy to be left to the community regarding venues, facilities, parks, and everything else designed for the Olympic Games (Rose & Spiegel, 2009). However, constant communication with all stakeholders must be maintained to develop an excellent strategic plan. In addition, planning should consider a time frame of ten to twenty years to move forward and adopt mixed management models between public and private companies[6]. For example, the city of Barcelona in 1976 had a General Metropolitan Plan[7] in agreement with the public administration for the reorganisation of the territory[8].

---

4   Rose, A. K., & Spiegel, M. M. (October 2009) The Olympic effect. *National Bureau of Economic Research*.

5   However, citizens being the central part of the strategic planning, the Organising Committee, without the support and involvement of the host community, will not be able to take advantage of all the intangible benefits associated with this type of "Mega-event".

6   PwC, Public-Private Partnerships: The US Perspective, June 2010.

7   Plan general metropolitano de ordenación urbana, de la entidad municipal metropolitana de Barcelona, Economic Study (February 1976), Corporacio Metropolitana de Barcelona (CMB), Barcelona, January 1976.

8   The urban project of the new Barcelona began in 1976 with the approval of the Metropolitan Plan, which included twenty-six municipalities considering them as

The following Metropolitan Plan ensured compliance with regulations, favoured the choice of Olympic areas, and significantly reduced the problems related to the choice of venues.

Thus, the case of Barcelona has become a model[9]. To follow in the management and organisation of these "mega-events". Unfortunately, not all states have the same capacities and human resources[10]. However, political, social and cultural conditions influence and condition the success of the Olympic event (della Sala, 2022). In this sense, we can say that not every city in the world can host these events and achieve the same results[11]. Each event should be analysed in its context and the historical moment in which it is organised. Only through post-Olympic planning will it be possible to establish a coherent emphasis on the city's image and territorial marketing. Cities such as Sydney[12], New York, Barcelona, Atlanta, and Sochi have invested heavily in marketing at all event stages: before, during, and after. There is no doubt that the most significant tangible value we can associate with a city's image is tourism (Preuss, 2000). For example, Barcelona has become one of Europe's leading cities in terms of the number of tourists, sporting events and international congresses. Its primary funding sources derived from the tourism sector are managed through joint ventures specialising in the different operations to be carried out. Moreover, Michael Pane, who has worked with the International Olympic Committee for more than twenty years, agrees that the optimal

---

part of total urban planning, using the communication and service roads as integration tools.

9   For example, Turin in 2006 and Rio in 2016 were two cities inspired by the Barcelona model for the exploitation of the Olympic event and have obtained entirely different results.

10 World Economic Forum, Global Risks 2020: A Global Risk Network Report, January 2010.

11 To host the two events (World Cup and Olympic Games), Brazil has received eighty-three billion in public and private funds for the development of new infrastructures, connections, communication, technology and other works that had to help the State solve some problems.

12 NSW Treasury: Office of Financial Management, The Economic Impact of the Sydney Olympic Games, November 1997. PwC, Business and Economic Benefits of the Sydney 2000 Games: A Collation of Evidence, 2001.

management of this type of "Mega-Event" should be entrusted to a joint venture with specialists in the economic sector[13]. Therefore, the bid for the Games can represent an opportunity for development, for a change in the operational philosophy of the host city[14]. However, the organising committees must consider all possible risks, present and future, always bearing in mind the importance of citizens and sustainable practices for the territory[15]. In addition, facilities planning must integrate urban social policies to support the city's development over time. Therefore, combining the event into a strategic plan can turn it into a long-term dynamic process. As stated by Gold (2008, 2016) and Roche (1992, 2002, 2003, 2006), mega-events, if successful in terms of organisation and promotion of the city's image, can contribute to the international projection of a new city's image and identity (Viehoff, 2018).

2.1.2. The impact of the Olympic Games

As noted above, from the first edition of the modern Olympic Games in Athens in 1896 until Tokyo in 2020, twenty-nine summer editions have been held, twenty-four of which have been organised in seventeen cities of different nations. While the first Winter Games were held in Chamonix in 1924 and until Beijing 2022, twenty-four editions have been collected in twenty-one cities of twelve countries. For this reason, observing different projects in different socio-economic contexts makes it possible to analyse the impact of the Olympic Games in other areas of the globe. Authors such as Andranovich and Burbank (2001), analysing urban impacts, identify spatial transformations as the most visible impact and one

---

13   Payne, M. (Spring 2007). *A gold-medal partnership*. Strategy+Business.
14   PwC, Cities of Opportunity, 2010.
15   Reassessing existing structures over time has become essential.

Table 7. Candidate cities for the Summer Olympic Games, 1896–2032
(Source: Compiled by author from IOC, 2022)

Year	Year of adjudication	Host city	Host country	Other candidates
1896	1894	Athens	Greece	London
1900	1894	Paris	France	
1904	1901	St. Louis*	United States	Chicago
1908	1904	London**	Great Britain	Berlin, Milan, Rome
1912	1909	Stockholm	Sweden	
1916	1912	Berlin	Germany	Berlin, Alexandria (Egypt), Budapest, Cleveland, Brussels
1920	1914	Antwerp	Belgium	Amsterdam, Atlanta, Brussels, Budapest, Budapest, Cleveland, Lyon, Havana, Philadelphia
1924	1921	Paris	France	Los Angeles, Atlantic City, Chicago, Pasadena, Rome, Barcelona, Amsterdam, Lyon
1928	1921	Amsterdam	The Netherlands	Los Angeles
1932	1923	Los Angeles	United States	
1936	1931	Berlin	Germany	Barcelona, Buenos Aires, Rome,
1940	1936	Tokyo	Japan	Tokyo, Helsinki, Rome
1944	1939	London	Great Britain	London, Athens, Budapest, Lausanne, Helsinki, Rome, Detroit
1948	1946	London	Great Britain	Baltimore, Lausanne, Los Angeles, Minneapolis, Philadelphia
1952	1947	Helsinki	Finland	Amsterdam, Chicago, Detroit, Detroit, Los Angeles, Minneapolis, Philadelphia

(*continued*)

Table 7. Continued

Year	Year of adjudication	Host city	Host country	Other candidates
1956	1949	Melbourne	Australia	Buenos Aires, Chicago, Detroit, Los Angeles, Los Angeles, Mexico City, Minneapolis, Montreal, Philadelphia
1960	1955	Roma	Italy	Budapest, Brussels, Detroit, Lausanne, Lausanne, Mexico City, Tokyo
1964	1959	Tokyo	Japan	Brussels, Detroit, Vienna
1968	1963	Mexico City	Mexico	Buenos Aires, Lyon, Detroit
1972	1966	Munich	Germany	Detroit, Madrid, Montreal
1976	1970	Montreal	Canada	Los Angeles, Moscow
1980	1974	Moscow	Soviet Union	Los Angeles
1984	1978	Los Angeles	United States	Tehran
1988	1981	Seoul	South Korea	Nagoya (Japan)
1992	1986	Barcelona	Spain	Amsterdam, Belgrade, Birmingham, Brisbane, Paris
1996	1990	Atlanta	USA	Athens, Belgrade, Manchester, Melbourne, Toronto
2000	1993	Sydney	Australia	Brasilia, Beijing, Berlin, Istanbul, Manchester, Milan, Tashkent
2004	1997	Athens	Greece	Buenos Aires, Cape Town, Istanbul, Lille, Rio de Janeiro, Rome, San Juan, St. Petersburg, Seville, Stockholm
2008	2001	Beijing	China	Bangkok, Cairo, Havana, Istanbul, Kuala Lumpur, Osaka SL, Paris SL, Seville, Toronto SL

Table 7. Continued

Year	Year of adjudication	Host city	Host country	Other candidates
2012	2005	London	United Kingdom	Istanbul, Havana, Leipzig, Paris SL, Madrid SL Moscow SL, New York SL, Rio de Janeiro
2016	2009	Rio de Janeiro	Brazil	Baku, Chicago SL, Doha, Madrid SL, Prague, Tokyo
2020	2013	Tokyo	Japan	Baku, Doha, Istanbul SL, Madrid SL
2024	2017	Paris	France	Unanimous
2028	2017	Los Angeles	United States	Unanimous
2032	2021	Brisbane	Australia	72 Yes, 5 No, 3 abstention (93,5% of valid votes)

of the most important legacies in the post-event phase[16]. Subsequently, authors such as Kasimati (2003, 2006) and Kassen-Noor (2013) identify infrastructure as the most perceptible and dangerous legacy for the future of candidate cities. Moreover, editions such as Rome, Tokyo, Mexico, Munich, Barcelona, Sydney, Turin, Vancouver and London allow us to observe how these mega-events remain an active and dynamic legacy of the host cities today (della Sala, 2022). Therefore, the metamorphosis of urban space and the transformation of streets and infrastructures imply new strategies to avoid compromising the future of citizens within the host cities[17]. Therefore, given the visibility of urban impacts and physical

---

16   Andranovich, G., Burbank, M. J., & Heying, C. H. (2001). Olympic cities: Lessons learned from the politics of mega-events. *Journal of Urban Affairs, 23*(2), 113–131. Arsen, D. (1997). Is there really a link between infrastructure and economic development? In R. D. Birmingham & R. Mier (Eds), *Dillemas of urban economic development: Issues in theory and practice* (pp. 82–98). Thousand Oaks, CA: Sage Publishing. Auruskeviciene, V., Pundziene, A., Skudiene, V., Gripsud, G., Nes, E. B., & Olsson, U. H. (2010). Change of attitudes and country image after hosting major sports events. *Inzinerine Ekonomika–Engeneering Economics, 21*(1), 53–59.

17   Bale, J., & Christensen, M. K. (2004). *Post-Olympic? Questioning sport in the twenty-first century.* Oxford: BERG.

Table 8. Candidate cities for the Olympic Winter Games, 1924–2026
(Source: Compiled by author from IOC, 2022)

Year	Year of adjudication.	Host city	Host country
1924	Chamonix	Francia	
1928	St. Moritz	Swiss	Davos, Engelberg (Switzerland)
1932	Lake Placid	United States	Montreal (Canada), Bear Mountain, Yosemite Valley, Lake Tahoe, Duluth, Minneapolis, Denver (USA)
1936	Garmisch-Partenkirchen	Germany	St Moritz (Switzerland)
1948	St Moritz	Swiss	Lake Placid (USA)
1952	Oslo	Norway	Cortina (Italy), Lake Placid (USA)
1956	Cortina	Italia	Colorado Springs, Lake Placid (USA), Montreal (Canada)
1960	Squaw Valley	EE.UU.	Innsbruck (Austria), St Moritz (Switzerland), Garmisch-Partenkirchen (Germany)
1964	Innsbruck	Austria	Calgary (Canada), Lahti/Are, (Sweden)
1968	Grenoble	Francia	Calgary (Canada), Lahti/Are (Sweden), Sapporo (Japan), Oslo (Norway), Lake Placid (USA)
1972	Sapporo	Japan	Banff (Canada), Lahti/Are (Sweden), Salt Lake City (USA),
1976	Innsbruck	Austria	Denver (USA), Sion (Switzerland), Tampere/Are (Finland), Vancouver (Canada)
1980	Lake Placid	EE.UU.	Vancouver-Garibaldi (Canada): withdrew before final vote
1984	Sarajevo	Yugoslavia	Sapporo (Japan), Falun/Goteborg (Sweden)
1988	Calgary	Canada	Falun (Sweden), Cortina (Italy)

Table 8. Continued

Year	Year of adjudication.	Host city	Host country
1992	Albertville	Francia	Anchorage (USA), Berchtesgaden (Germany), Cortina (Italy), Lillehammer (Norway), Falun (Sweden), Sofia (Bulgaria)
1994	Lillehammer	Norway	Anchorage (USA), Oestersund/Are (Sweden), Sofia (Bulgaria)
1998	Nagano	Japan	Aoste (Italy), Jaca (Spain), Oestersund (Sweden), Salt Lake City (USA)
2002	Salt Lake City	EE.UU.	Oestersund (Sweden), Quebec City (Canada), Sion (Switzerland)
2006	Turin	Italia	Helsinki (Finland), Klagenfurt (Austria), Poprad-Tatry (Slovakia), Sion (Switzerland), Zakopane (Poland)
2010	Vancouver	Canada	PyeongChang (South Korea), Salzburg (Austria)
2014	Sochi	Russia	PyeongChang (South Korea), Salzburg (Austria)
2018	Pyeongchang	Korea del Sur	Annecy (France), Munich (Germany)
2022	Beijing	China	Almaty (Kazakhstan)
2026	Milan-Cortina	Italia	Stockholm

transformations in Olympic cities, the first classification related to urban intensity is observed through the contribution of Essex and Chalkley (1998):

- Low impact: Athens (1896), Paris (1900), St Louis (1904), London (1948), Mexico (1968), Los Angeles (1984).
- Games that have focused on the development of sports facilities: London (1908), Stockholm (1912), Los Angeles (1932), Berlin (1936), Helsinki (1952), Melbourne (1956), Atlanta (1996).

Table 9. Urban impact of the 2004–2028 Summer Olympics (Source: The following elaboration was provided from the groups provided by Essex and Chalkley in 1998)

Low impact	Sports facilities	Urban transformation
Paris 2024 *	Beijing 2008	London 2012
Los Angeles 2028**	Rio 2016	Athens 2004
		Tokyo 2020

* The Paris 2024 edition will have 95 per cent temporary or existing structures.
** The 2028 edition of Los Angeles will be an event with 100 per cent temporary or existing structures.

- Games that have transformed the city's urban identity: Rome (1960), Tokyo (1964), Munich (1972), Montreal (1976), Moscow (1980), Seoul (1988), Barcelona (1992), Sydney (2000).

However, the following groups only refer to the summer editions up to 2000. In the following table, the Olympic Games that took place until today and could not be observed then were added. Next, the table (Table 10) shows a classification of the urban impact of the winter edition.

On the other hand, winter editions require transformations of mountain sites, specific facilities and finally, a revolution in the transport system, which shows that they developed differently from summer editions over time.

The following table shows the different editions divided into three groups according to the impact generated.

As we will observe later, urban transformation and the design of spaces, over time, gain enormous importance in the social and economic aspects of the city. The planning and construction new sports facilities in mountainous areas is a sensitive issue, especially considering the natural environment and landscape. In addition, ski jumping and bobsleighing facilities are two of the most problematic facilities, which continue to raise doubts and criticism from the IOC. However, the space transformation should be integrated into a dynamic structure rooted in a long-term plan (della Sala, 2022). The authors Chalkley and Essex, in agreement with Preuss (Preuss, 2004), emphasise the importance of effective facility design in the

Table 10. Urban impact of the Winter Olympics 1924–2026 (Source: Own implementation)

Low impact	Sports facilities	Urban transformation
Chamonix 1924	Cortina 1956	Oslo 1952
Saint Moritz 1928	Squaw Valley 1960	Innsbruck 1964
Lake placid 1932	Lake Placid 1980	Grenoble 1968
Garnish 1936	Sarajevo 1984	Sapporo 1972
Saint Moritz 1948	Lillehamer 1994	Innsbruck 1976
Calgary 1988	PyeongChang 2018	Albertville 1992
Salt Lake 2002		Nagano 1998
Milano-Cortina 2026		Turin 2006
		Vancouver 2010
		Sochi 2014
		Beijing 2022

post-Olympic period[18], which tends to encourage the evolution of sporting practices and provide accommodation for poorer citizens (Chalkley, 1999).

Subsequently, Hiller (2014), looking at mega-events in the urban process, has identified the following phenomena that can be realised in candidate cities:

- The catalyst for urban change.
- Land-use change in urban space.
- Stimulation of creativity in spatial planning.
- They mobilise funding (private and public).
- Support in projects is considered to be very ambitious or costly.
- Requires completion by the date of the event.
- Structural improvements in specific sectors (e.g. transport).
- It produces specific structures that redefine urban space and territory.

18   Chalkley, B., & Essex, S. (1999). Urban development through hosting international events: A history of Olympic Games. *Planning Perspectives, 14*, 369–394.

However, urban transformations can have different impacts in various fields: socio-cultural, political, economic, sporting, medical, infrastructural, etc. Moreover, since Turin 2006, environmental impact has become a key element in the election of candidate cities. Over time, transforming the city's image through promoting a healthy lifestyle can contribute to increasing national pride and having a socio-cultural impact on the host community in the future. Furthermore, the international promotion of the Olympic city should motivate businesses and investors to visit the city and take advantage of the new services developed for the post-Olympic future (Billings, 2012). For example, Barcelona's post-Olympic planning was based on the organisation of new areas specifically for technological development and therefore, telecommunication investments were included in the budget to offer companies new high-tech services (Brunet, 2005). While in Sydney, pre-Olympic planning was an example of a promotional campaign for international companies. The city and State conducted targeted campaigns to encourage global companies to hold conferences and events in the city that hosted the 2000 Olympics. In addition, promotional activities enabled the city of Sydney to host continuous events over four years. The most important result was Sydney's inclusion in the international conference, convention and event market, which can be described as surprising or unprecedented[19]. Therefore, the Olympic Games can only guarantee a unique development if the quality of management and planning borders on perfection (Chalkley, 1999; Essex, 1998; Gratton, 2001; Preuss, 2000, 2004).

Indeed, one question that arises concerns the motivation of cities to host such a large and complex event. Therefore, why do cities want to host the Olympics? Over time, we have observed various political motivations that remain vital to host states. In addition, in recent years, we have witnessed a complete transformation of the bidding process and public

---

19   To achieve the following result, the Sydney Organising Committee involved the top experts in Olympic planning, ensuring a unique development for the entire community.

presentations of each host country by prime ministers in favour of the bid[20]. Thus, at the political level[21], the Olympic event has often been presented as a way to encourage the creation of new jobs and to improve each country's gross domestic product[22]. However, since the early 2000s, protest groups[23] against the organisation of the Olympic event have increased dramatically to become active movements[24]. That has forced public administrations to withdraw political bids (Heine, 2018). Therefore, the referendum phenomenon remains a fundamental element for recognising and affirming a shared development model among all *event stakeholders*.

On the other hand, the economic impact resulting from the Olympic Games is highly controversial. Some authors have conducted in-depth studies on financing structures, others on public capital investments, still others on marketing, and others on the organisation of *stakeholders* strategically for the future of the candidate cities.

For this reason, it is currently rather difficult to quantify the real economic benefits for candidate cities. For example, Preuss, in his 2004 study, identifies tourism as the maximum financial benefit that Olympic cities can derive over time[25].

20 In 2021, for the first time, Australia's Prime Minister presented the official Melbourne 2032 bid eleven years before the Olympic event, something that had never happened before.

21 Political interest focuses on the possibility of attracting new foreign investors and increasing the capital available to meet or attempt to meet the real needs of citizens.

22 Matheson, V. A. (2006). *Mega-events: The effect of the world's most significant sporting events on local, regional, and national economies.* Worchester: College of the Holy Cross, Department of Economics. Matheson, V. A., & Baade, R. A. (2004). Mega-sporting events in developing nations: Playing the way to prosperity? *South African Journal of Economics, 72*(5), 1084–1095. McDonogh, G. (1991). Discourses of the city: Policy and response in post-transitional Barcelona. *City and Society, 5*(1), 40–63.

23 For example, Munich in 2018 was forced to withdraw its bid because citizens, through a popular referendum, did not want any events in those locations chosen by third parties.

24 Nolympics is recognised as one of the most active movements internationally.

25 Preuss, H. (2004). *The economics of staging the Olympics: A comparison of the games, 1972–2008.* Cheltenham, UK: Edward Elgar Publishing.

In addition, Preuss, in 2000[26] identifies two critical points in the consideration of economic aspects:

- *Increased demand* Increased the number of employees, increased the domestic economy, and increased employees and profits. Preuss foresees rapid inflation after the Olympics as a possible downside if the economic restructuring plan is not planned or does not work.
- *International attractiveness→* Increased numbers of tourists who, if they leave positive feedback, can contribute to incalculable long-term benefits. In addition, the increase in tourists and travellers will enable the city to acquire a new image in the global community.

Agreeing with Preuss, the economic analysis of the event should be broken down into spectator spending associated with the event and indirect non-Olympic spending. Furthermore, other elements, such as the level of unemployment in Olympic cities, should not be considered and compared to the construction period of the Olympics. Logically, unemployment decreases during the Olympic construction period due to practical needs, which will be different in the post-Olympic period. Furthermore, Preuss (2000) identifies the price increase[27]. As the most significant negative impact associated with increased tourism in the candidate cities, rising prices can quickly lead to temporary inflation in the host community, which will inevitably imply increased social inequalities. Finally, transforming the city's image and building a city brand can help attract millions of tourists and businesses willing to invest in the host city over time. However, the increased flow of tourists and business pressure on the city may undermine the development of the host society, implying new inequalities over time (Smith, 2009).

In conclusion, the physical transformation of the city and its image is a delicate issue that must be planned dynamically between the city and its citizens. In cities such as Barcelona, for example, tourism has become

26   Preuss, H. (2000). *Economics of the Olympic Games*. Sydney: Walla Walla Press.
27   For example, in Barcelona in 1993, one year after the event, prices increased by 240 per cent compared to the pre-Olympic period.

a fundamental element of the entire region's economy and, over the years, has continually evolved at the expense of the citizens. For example, in Barcelona, the loss of symbolic places, the transformation of new areas, the provision of new offices, the increase in prices, and the design of unique hotels have caused the expulsion of citizens from the central area to the periphery. Moreover, rising prices and job insecurity are considered the main consequences of the gentrification of Barcelona. Like Barcelona, London in 2012, through its Olympic bid, caused the gentrification of the Olympic area due to rising prices and changes in the labour market. In recent years, studies on the impact of the Olympic Games have increasingly focused on exploiting new mixed models for post-Olympic management and organisation.

2.2. Mega-events and territorial transformations

2.2.1. Urban transformation at the Summer Olympics

Over time, mega-events have induced major urban transformations in host cities. As we will observe in the next chapter, mega-events have developed and evolved through a model of sports promotion that has become a model of metropolitan development over time. Therefore, to analyse the different main phases of urban transformation in the host cities, we will examine the evolution through five phases.

2.2.1.1 Phase I: Minimal transformation (1896–1924)

The first phase of the Olympic Games began with the event's first edition and lasted until the first Olympic Village in Paris was constructed in 1924. The following editions are characterised by private funding to promote the sport through host cities and economic organisations. Therefore, in this first phase, Olympic cities will only carry out minimal transformations, proposing a temporary accommodation model in military areas or through public availability[28].

2.2.1.2 Phase II: Emerging spatial organisation (1932–1956)

Later, in the second phase, the Olympic cities would focus mainly on constructing sports facilities to establish a new sports district on the outskirts of the cities. The second phase began with Los Angeles in 1932 and lasted until Melbourne in 1956.

Hence, the Olympic editions in the following phase introduced and fostered the construction of new sports facilities and Olympic housing that would constitute the new neighbourhoods for the host cities. The second phase, therefore, saw the emergence of spatial organisation and the creation of a basis for the infrastructural transformations that we will observe in the third phase. As we will follow in the next chapter, the Berlin 1936 project will stimulate the future organising committees of Helsinki and Melbourne to build a new sports district in a suburban area of the city.

2.2.1.3 Phase III: Reconfiguration of cities (1960–1988)

In the successive phase, the Olympic cities will be deeply inspired by the 1960 Rome Olympic Games project. The Rome edition is the first to consider the Olympic event as a tool for urban development and as an opportunity for reconfiguring the city (della Sala, 2022).

Rome concentrated on different areas, developing a modern transport system, including constructing the airport. Subsequently, the 1964 Tokyo edition followed the same philosophy as the previous edition, using the Olympic event as an instrument of urban renewal. However, Tokyo took advantage of the event to promote a ten-year development plan that included improving the infrastructure system, roads, harbour, housing, water supply and public health. The 1964 Tokyo edition is one of Olympic history's most extensive urban development projects.

---

28 In St Louis in 1904, the Games were held over several months as an adjunct to the World's Fair and the swimming events were held in an artificial pool at the fairgrounds (Gordon, 1983).

For the 1968 Mexico City edition, a spatial organisation was planned that included the development of new infrastructure and housing to expand a peripheral area of the metropolis. Meanwhile, the Munich 1972 edition proposed redeveloping a brownfield site to construct a sports park, including residences[29].

The Munich Plan included the construction of a new self-sufficient community and other infrastructural improvements in the Bavarian capital. In addition, various improvements were made, such as the restoration and pedestrianisation of the historic city centre, the expansion of public transport lines, the creation of underground car parks, the development of a new shopping centre, and the construction of three new motorways. Subsequently, the Montreal 1976 and Moscow 1980 editions proposed new housing and infrastructure works for the reconfiguration and modernisation of the cities. However, the Montreal edition is recognised as one of the most worrying moments for the increase in the size of the Olympic event. Finally, the 1984 Los Angeles Games will be organised through private funding with existing or temporary facilities[30]. On the other hand, the 1988 edition in Seoul allowed the Olympic Games to resume their role as a vehicle for urban transformation. The Seoul project was based on a twenty-year plan that introduced new programmes to ensure higher health and hygiene standards throughout the city. In addition, the project included measures for air pollution, waste control, water quality and a significant decontamination plan for the Han River. Thanks to the Olympics, Korean cities were able to develop three new underground lines and expand forty-seven bus lines to relieve traffic congestion. Finally, the airport was raised, and new projects were designed to emphasise the cultural aspects of the Olympic event. The city was able to have a programme of renovation and reconstruction of historical monuments, such as palaces and shrines.

29   To improve orientation, they were coloured to match the separate areas or residential places to which they led and were also equipped with signs, notice boards or display boards (COJO, 1974, p. 109).

30   To avoid large capital expenditures, the organisers used existing sports facilities and accommodation over a wide geographical area, including the Olympic Stadium in 1932 and student residences at the Universities of California and Southern California (Essex, 1998).

2.2.1.4 Phase IV: Large-scale urban transformations (1992–2004)

The fourth phase will begin in Barcelona in 1992 and end in Athens in 2004. The 1992 Catalan edition is recognised as the best example of the role of the Olympic Games as a catalyst for change and urban renewal[31]. Meanwhile, the 2004 Greek edition will leave many doubts and signify a new crisis for Olympism.

Barcelona, in 1992, introduced a new strategy for the reconstruction and redefinition of post-industrial cities. The transformation of a city in crisis will be the common element of all the Olympic cities in the following phase. Furthermore, Barcelona has become an example of post-industrial reconversion by constructing a new image for exploiting tourism in the post-Olympic period. In this way, the Olympic Games became a means to ensure major transformations in urban infrastructure through a mixed economy. From 1992 onwards, tourism became a vital element of the host city economy in the post-Olympic phase (Hughes, 1993). Barcelona will present a new image and development strategy for candidate cities (Moragas, 1995).

Following the same philosophy as Barcelona, in 2000, Sydney proposed an ambitious project to reconfigure brownfield sites by applying new sustainable practices. Sydney is recognised as the first Olympic city to introduce the theme of environmental sustainability in developing the Olympic event. The stadium and Olympic Village have been located in the *Homebush Bay* area[32], located in a suburban area of the city. The area had been neglected for many years, it was only thanks to Olympia's bid that the municipality strengthened and accelerated the redevelopment of the entire area, establishing a new structural plan for reconfiguring it. Subsequently

---

31   By the end of the Olympic event, 330 urban interventions were counted (Holsa, 1992).

32   The Olympic Stadium and Village will be in an area known as Homebush Bay, which is 14 km west of the city centre (Young, 1992; NSW Government, 1994; Brogan, 1996). The 760-hectare site has been marked for many decades by noxious land uses and areas of contaminated brownfield land used for dumping domestic and industrial waste, including State Brickworks, State Abattoir and Royal Australian Armaments Depot (Sanders, 1995; Essex, 1998).

2004, the Athens project was included in a programme to transform the Greek city's primary infrastructure[33]. The reconfiguration of the harbour, the redevelopment of the central areas, the construction of a new airport and the provision of the metro system were the main advanced infrastructure works for the modernisation of Athens (Georgiadis, 2016).

2.2.1.5 Phase V: Metropolitan development (2008–2028)

Finally, phase five is considered from Beijing 2008 to Los Angeles 2028.

In this phase, the cities are characterised by metropolitan development that uses the central empty spaces to reconfigure the host cities. Thus, the Olympic Games will place greater emphasis on environmental protection and the sustainable development of the Olympic project. The establishment of an environmental park[34] In Beijing (Jia, 2012), planning a water recycling system in London can be considered an innovative measure for environmental protection in the candidate cities. Furthermore, since London 2012, the tangible and intangible Olympic legacy has assumed great importance for post-Olympic planning (Imrie, 2008). As such, temporary facilities and innovative solutions will be used in the following editions.

The Beijing[35] Compared to previous phases, London projects have favoured the emergence of one-off infrastructure works such as airport reconfiguration and rail and metro expansion.

Subsequently, the 2016 edition of Rio de Janeiro will bring further changes in the allocation of host cities, which, for the first time, will be chosen without any competition. The event's awarding through a proclamation process will imply including temporary structures and reusing

---

33 The Games master plan was described by Beriators and Gospodini (2004, p. 197) as a dispersed model that suggests a strategy to promote regeneration and multi-core urban development.

34 At the end of 2008, Beijing had a total expenditure of $12.2 billion for protecting and improving the ecological environment. The overall project foresees the construction of fourteen sewage treatment plants for water treatment from 42 per cent to 905 (COJO, 2010).

35 Beijing is a city that has undergone a significant transformation in recent decades, rapidly metamorphosing into an internationalised metropolis (Cook, 2006).

existing sports facilities in the candidate cities. Therefore, the IOC has identified Paris and Los Angeles as cities that could represent the new evolution and organisation of the Olympic event. The following steps allow us to state that the Olympic Games throughout urban history have been a source of inspiration for candidate cities and that the specific variables of each municipality have favoured the legitimisation of the Olympic city as a distinct urban genre. Moreover, urban planners have proposed different projects that have become development models for other cities over time. The following steps help us reflect on the history of the physical impact of the Games and how it has changed over two centuries.

Table 11. Summary of the main urban transformations of the Summer Olympics
(Source: Own implementation on Essex, 1998)

Phase I	1896-1924	Minimal transformation. Sports cities.	Prospects for the development of sport. Temporary accommodation. The event was initiated by private interests but jointly funded by the public sector.
Phase II	1932-1956	Emerging spatial organisation Emerging cities	Creation of a sports quarter in the peripheral areas of the cities Sports facilities The foundations for infrastructure development are laid.
Phase III	1960-1988	Urban development tools, especially transport, Olympic Villages and economic development opportunities Reconfiguration of cities	Infrastructure investment as part of modernisation Public sector financing Concern arises over the increasing size of the event

Table 11. Continued

Phase	Years		
Phase IV	1992-2004	Large-scale urban transformations. Post-industrial development tool Cities in crisis	The role of the Olympic Games as a means of securing a significant change in urban infrastructures Mixed economy due to new marketing revenues from the event The Olympic Games as a tool for promoting a new image of the candidate cities Tourism asserts itself as a vital element of the Olympic cities' economy.
Phase V[36]	2008-2028	Metropolitan development and legacy planning Global cities	Metropolitan development in empty spaces. Tool for the redefinition and reorganisation of the urban fabric. Greater emphasis on environmental protection and sustainable development. The Olympic legacy assumes great importance in the planning of the post-Olympic phase. Temporary installations and demountable solutions

2.2.2. *Urban transformations at the Winter Olympics*

The first Winter Olympic Games were held only thirty years after the restoration of the modern Olympic Games. Therefore, winter sports were excluded from the original conception of the Olympics. However, the

36 The next phase was implemented by researching the Olympic events held up to the date of the study.

popularity of winter sports and the creation of tourist sites in mountain areas favoured the establishment of a winter edition. Therefore, the IOC allowed a week of winter competitions to be organised six months before the 1924 Paris Olympics.

2.2.2.1 Phase I: Promotion of sports tourism (1924–1932)

The event's first edition was held in 1924 in Chamonix, and until 1948, the cities awarded the summer event and had the option of organising the winter edition.

Therefore, the first phase[37] began with Chamonix in 1924 and lasted until the 1932 edition in Lake Placid[38]. The first editions are characterised by the limited size of the event site and the development of sports tourism by providing new sports infrastructures. In the later phase, projects were financed by private interests or included in regional development to establish tourist sites. Meanwhile, the accommodation of athletes was based on the availability of hotels in each mountain resort. The construction or renovation of sports facilities has been a requirement of host venues throughout the Olympics. Therefore, the early period is characterised by mountain resorts interested in winter sports tourism.

2.2.2.2 Phase II: Development of sports facilities (1936–1960)

The second phase has the first characteristics: small mountain sites, minimal infrastructure investment and public participation in a winter sports tourism development project. The only exception in this phase is Oslo, which in 1952 proposed a major urban renewal and reconfiguration project, building for the first time three Olympic Villages in the city[39]. Oslo is the only city with an almost half a million-resident population. However, the increase in the size of the event and the number of participants began

---

37 During this phase, the participating athletes ranged from 250 to 500.
38 The 1932 Winter Olympic Games were held in Lake Placid and involved three new constructions and two substantial renovations of existing facilities (Essex, 2007).
39 The infrastructure investment plan for the Oslo Games was controversial for a city still emerging from severe post-war austerity.

to generate new infrastructure needs. Therefore, Oslo's infrastructure investments are beginning to form a form of infrastructure development in the post-Olympic period.

2.2.2.3 Phase III: Tool for regional development (1964–1980)

The third phase is characterised by a series of definitive changes:

- An increase in the number of athletes.
- Larger Olympic cities.
- Regional development.

The next phase begins with the Innsbruck edition in 1964 and temporarily extends until the Lake Placid edition in 1980. The winter edition projects then become an instrument for transforming the regional transport system of the host countries. In the next phase, governments will become key players in financing the Olympic event. The 1964 Innsbruck edition[40] favoured the construction of a new district to house the Olympic athletes and new sports facilities to promote winter sports and equipment. Meanwhile, the 1968 Grenoble project was one of the most significant infrastructure projects realised in the winter edition. Thus, the modernisation process of the entire region was accelerated by the Grenoble edition, which ensured substantial infrastructure investments and better connections to the French capital through new railway lines, two airports and motorways. The reconstruction of roads accounted for 20 per cent of the total investment for Grenoble (COJO, 1969, p. 46). In addition, the investment included a link between Grenoble and Geneva that was a catalyst for the regional economy, turning the city into an important university and conference centre[41].

Subsequently, in 1972, Sapporo proposed an infrastructure renewal project that followed the Japanese government's plan to improve the

---

40 During this phase, television revenues became an essential funding source (see Essex, 2007).
41 the Grenoble Olympic Village was built in a Priority Urbanisation Zone (ZUP).

country's infrastructure system. The infrastructure investment included the expansion of two airports, the upgrading of the central railway station, 213 km of roads and a 45 km rapid transit system[42]. The Sapporo event allowed the IOC to reflect on the growing scale of Olympic projects and the risk of accumulating debts to provide infrastructure. Meanwhile, the 1976 Innsbruck event only had to give a new Olympic Village, as the blocks built in 1964 were converted into residences and were no longer available. In the next phase, the Olympic Village became a central urban element in transforming an essential venue for the event.

2.2.2.4 Phase IV: Large-scale transformations (1984[43]–1998)

The next stage is therefore characterised by an increase in infrastructure works and a new role for the winter event in the context of regional transformation.

Furthermore, the fourth phase is characterised by the construction of accommodation for athletes, audio-visual rights, and the increase in spectators, which has led to increased infrastructure work for transport. While, since 1988[44] Olympic accommodations required two or more Olympic Villages to house the athletes. In the next phase, ending with the 1998 Nagano edition, the demands related to the size of the event meant that larger cities with larger populations had to be chosen. Therefore, the role of the winter edition intensified and became a means to ensure infrastructural transformation and modernisation in the host cities. The next phase, however, will see bidders such as Calgary and Lillehammer providing two

42   The Japanese government saw the 1972 Sapporo Games as a unique economic opportunity to invigorate the island of Hokkaido. Only 5 per cent of the initial capital was used to construct sports facilities. Ninety-five per cent of the capital was used for infrastructural improvements.

43   For the 1984 Sarajevo Games, the citizens of the region agreed to make a voluntary contribution from their monthly salary to an Olympic Development fund between 1982 and 1984 (for Bosnia and Herzegovina, 0.2 per cent and Sarajevo itself, 0.3 per cent (Essex, 2007).

44   In Calgary, the event's hosting led to overestimating some sports facilities, such as the *Olympic Saddledome*.

projects to revive the local economy (Spilling, 1998). One of the significant regional winter sports promotion projects is the 1992 Albertville edition, which renovated a small site with spas for the construction of the temporary Olympic Village[45]. However, the Organising Committee proposed other locations with existing hotel accommodation. The Lillehammer project promoted the construction of a temporary Olympic Village consisting of 200 wooden chalets that could be easily dismantled and dismantled in the post-Olympic period. The city aimed to position itself as a venue for sporting events while following the trajectory observed in the previous stage. Lillehammer was the first edition to incorporate sustainable development objectives. Meanwhile, Nagano proposed a project with sustainable development principles and infrastructural works to renew the regional fabric[46]. (Nakamura, 2017). The following phase was characterised by increased infrastructure work and an intensification of the Olympic accommodation offer.

2.2.2.5 Phase V: Sustainable development and regional planning (2002–2014)

During the fifth phase, Olympic projects continued to require large infrastructural investment but with a greater focus on environmental protection and sustainable development. In addition, the 2010 Vancouver Games[47] Included an Olympic legacy plan for the first time, marking the importance of strategic planning at the Winter Olympics. The following phase consists of the Olympic editions up to Sochi 2014. As we will observe later, large metropolitan cities and neighbouring or surrounding mountain resorts began to organise winter projects, becoming a multi-event within a larger space. The 2006 edition in Turin is recognised as

45 The dispersion of the Olympic site in Albertville led to some reflections on the Olympic Village as a place of experience and cultural exchange.

46 After 1988, planning multiple Olympic Villages to accommodate athletes closer to the sports venues became necessary. From then on, media accommodation was a mandatory challenge to meet Olympic requirements.

47 The Vancouver project was part of the transformation of a disused waterfront that needed urban planning and regeneration to fit into the fabric of the metropolis.

the first metropolis to involve an Olympic area comprising seven mountain resorts and three Olympic villages. In addition, the assignment of the
2006 Olympics to Turin will increase the size of the winter event, which
will be hosted by cities with more than one million inhabitants. The event
is part of a strategy of transformation and redefinition of metropolitan
areas abandoned by the industrial crisis of the 1980s[48].

Therefore, the Turin 2006 project can be compared to that of Barcelona
1992 in that it incorporated the event into a broader urban transformation
strategy with a solid foundation supported by master and long-term strategic plans. The city of Turin, thanks to the inclusion of the Olympic event,
had the opportunity to develop new areas, restore the old railway line, add
new mixed-use facilities, restore pedestrian areas in the historic centre and
other works, including the revaluation of emblematic spaces and buildings.
Turin 2006 will enable the IOC to promote a new image for the event and
a unique opportunity for the host cities[49].

Meanwhile, Vancouver 2010 focused on the realisation of the sports
facilities and on reducing the impact of the Olympic event through new
measures for sustainable development[50].

Vancouver, including participation, volunteering and a foundation
for Olympic legacy management, provided a new model for managing and
organising the post-Olympic period over time (VanWynsberghe, 2012).

On the other hand, Sochi 2014[51] Sochi was an ambitious project to
build a mountain tourism venue in a summer resort area of the former
Soviet Union. However, the cost associated with the event was less than

---

48   The city of Turin, in 1995, thanks to the approval of the Gregotti-Cagnardi Master
      Plan, started the transformation of its urban structures through the three guidelines of the "*Le Spine*" plan. The plan was intended to improve transport access and
      to reuse areas abandoned by the industrial past.

49   For the implementation of the Olympic works, the TOROC adopted an environmental assessment system for the evolution of the impacts on the territory. The
      environmental management system was awarded ISO14001 status.

50   The Vancouver event strongly emphasised its planning tools through a philosophy
      of long-term sustainable development.

51   Awarding the 2014 Winter Olympics to Sochi in Russia may represent the beginning
      of a new phase or even a step backwards in the event's trajectory (Chappelet, 2008).

50 billion euros, leading to concern throughout the Olympic community about the high costs of organising the event[52].

The Sochi edition was controversial in many respects, especially during the staging and the choice of the ground for the construction of the Olympic infrastructure.

2.2.2.6 Phase VI: "The Future" (2018–onwards)

Finally, the sixth phase, the future of the Winter Olympics, is developing thanks to a new model of Olympic regionalisation through the constitution of a metropolitan centre and other mountain sites adjudicating the Olympic event and competitions.

Therefore, the winter edition continues to be transformed through a metropolitan and regional development model in different metropolises around the world. Beijing 2022 has provided a regional model with two mountain towns connected by a new infrastructure network developed to promote the new national sports strategy. The host city of Beijing and the village of Zhangjiakou are about 231 km apart. The following organisational model allows us to observe how the dimension of the winter edition continues to expand and be organised in one or more territories. Furthermore, in consideration of the nomination of the Milan-Cortina 2026 edition[53], the inclusion of a project comprising two Olympic cities and three regions of the host country will be observed for the first time. Therefore, future projects will allow for a bid with two states willing to collaborate to expand trade and economy and create possibilities of synergy between the host states.

---

52   Seven power plants (some thermal and hydro) were built or renovated to increase the capacity of the region's power grid, ensuring a stable power supply for the event (SOOC, 2009).

53   The two Olympic cities are not directly connected by an infrastructural system.

Table 12. Summary of the main urban transformations of the Olympic Winter Games (Source: Essex & de Groot, 2016)

Phase I	1924-1932	Minimal infrastructural transformation, apart from sports facilities Small host populations (around 3,000)	Development prospects for winter sport tourism Winterisation' of existing accommodation rather than over-provision of accommodation The event was initiated by private interests but funded jointly with the public sector. Environmental concerns raised
Phase II	1936-1960	Emerging infrastructural demands, especially transportation Small host populations (around 13,000)	Growing volumes of participants and spectators requiring investment in transportation The limited basis for other permanent infrastructure
Phase III	1964-1980	Tool of regional development, especially transportation, Olympic Villages and economic development opportunities Medium host populations (around 100,000 or more)	Infrastructure investment as part of regional modernisation and development Substantial public sector funding but emerging television revenues Concern emerges about the increasing size of the event about camaraderie, transport problems, debt and environmental damage

Table 12. Continued

Phase IV	1984-1998	Large-scale urban transformations, including multiple Olympic Villages Large host populations (c. 300,000)	Role of the Winter Olympics as a means to secure major urban infrastructural change Higher television revenues More formal recognition of environmental issues in planning and development
Phase V	2002-2014	Sustainable development and legacy planning Large populations/ metropolitan areas (c. 1 million)	Large-scale infrastructural investment and redevelopment reflecting global and political ambitions of host cities High television revenues Greater emphasis on environmental protection and sustainable development through environmental management systems The emergence of 'soft' legacies (social inclusion, human rights, integration of indigenous cultures), but also more international debate and controversy over the adoption of international norms
The 'Future'	2018— onwards	Reluctance to host Olympic Games in developed economies related to scale, cost, and demands of the event	More interest from emerging economies with centralised governments wishing to achieve political acceptance on a global stage Reform of IOC expectations for the event: hard and soft legacies determined by dialogue rather than pre-defined; relaxation of 'compact games' concept

2.3. The temporary transformation of the transport system

2.3.1. The Olympic transport system

Throughout the history of the modern Olympic Games, the management of the transport system has evolved to be considered an essential element of the event's overall success. The ability to effectively utilise the extraordinary investment in transport has become a key element in creating permanent infrastructure and changing the transport system of Olympic cities. Until the end of the twentieth century, transport was seen as a necessity for linking the main Olympic venues (della Sala, 2022). However, the increase in the spatial dimension of the event, the number of participants and visitors, implied a redefinition of the transport system within the Olympic cities. Therefore, Olympic towns have started using the event to change travel patterns in the most congested metropolises (Kassen-Noor, 2016). Transport management and modes are one of the main elements of Olympic funding.

Furthermore, the transport system is highly complex to plan. Although little research has been conducted on Olympic transport, there are enormous benefits of reorganising the road system that can become a catalyst for host cities in the post-Olympic period. As we will see in the next chapter, the spatial dimension of the Winter Games implies an enormous infrastructural connection for the Olympic venues. Meanwhile, the summer edition indicates shorter distances and an increased flow of people during the weeks of the event. Therefore, both editions allow for the realisation of various projects that can bring enormous long-term benefits to the territory and the host community.

For this reason, the planning of the summer edition is more complex and requires more time and resources in a limited amount of time. However, the proximity of the Olympic venues reduces travel time, encouraging other types of mobility. At the other extreme, the Winter Games are in mountainous areas and tend to require more vehicles and travel time, implying a redefinition of the entire infrastructure system. This difference between the two events increases the probability of non-use and failure of the infrastructure system planned for the winter event (della Sala, 2022). Furthermore, we must pay attention to the event development for Paralympic athletes. However, research on adapted and adequate transport is still in its infancy and mainly focuses on the accessibility of venues, including vehicles such as buses and minibuses and transport stations. The critical moment when the operation of Olympic transport was reorganised was in Atlanta in 1996 (Kassen-Noor, 2017). The congestion in the city during the event led to a revision of the transport system by the IOC to ensure that every athlete would be on time for competitions. Henceforth, future Olympic venues will need specific and priority routes within the host cities (Batuhan, 1996). Secondly, a detailed and sophisticated body will manage the Olympic transportation system (OTS) during the Olympic event. Third, the media should be accredited to receive an exclusive public service, ensuring timely and reliable broadcasting of the competitions (Kassens-Noor, 2013). Thus, in Nagano in 1998, Japan proposed a new traffic management system (UTMS) to reduce temporary congestion (Tanaka, 1997). In addition, Nagano improved and expanded the local road and highway network.

Meanwhile, Sydney 2000 provided a new railway line and special buses for Olympic visitors. The Salt Lake 2002 Organising Committee included Olympic transport in the long-term goals of the city and region. In this way, transport began to be seen as a tool for improvement and a tangible benefit for the area. In 2004, Athens demonstrated a remarkable ability to rethink the transport system for the Olympic event in consideration of metropolitan development plans. The extension of the metro, train and airport were some of the projects realised for the Olympic event and became tangible elements in the post-Olympic period. In addition 2006, within the total budget, Turin included 800 million in specific funds for infrastructure improvements. The construction of new roads, motorways, underground railways, and other routes around the metropolitan city enabled the city of Turin to improve transport in the post-Olympic period.

Meanwhile, the Vancouver 2010 Games ensured a new railway line was constructed connecting the airport and downtown Vancouver (Bracewell, 2009). Subsequently, the London 2012 Games allowed the city to build a new railway station in an area previously only connected by the underground. Meanwhile, in Sochi 2014, connecting three Olympic venues in an area that had yet to be exploited led to a considerable effort to build Olympic infrastructure. On the other hand, in Rio 2016, the transport system is considered the most important legacy of the area. The Bus Rapid Transit (BTR) service inspired other Olympic cities by connecting the Olympic area with four different city areas. Finally, Tokyo 2020 has again upgraded its high-speed railway lines, reinforcing its reputation as a world leader in this technology. The Intelligent Transport System (ITS) was the cornerstone of Tokyo's bid (Kassen-Noor, 2017).

2.3.2. *Planning the Olympic transport system*

The main rule for creating the Olympic transport system is to integrate it into each country's strategic transport system planning. However, peak transport demand during the Olympic event will result in temporary changes. The modifications can be used to analyse the peak flow of the transport system (della Sala, 2022). After the results of Atlanta in 1996, Olympic transport planning became an increasingly complex challenge for host cities. Over the years, the IOC has established minimum criteria to ensure smooth traffic conditions by providing a primary transport system. Therefore, the provision of preferential lines will allow specific configurations to be proposed for each of the Olympic venues. However, the IOC's general recommendation only concerns minimising travel time between venues and Olympic Villages. Following the 2018 reforms, the IOC supports the Organising Committee in researching or implementing new solutions for the Olympic transport system. For example, for the 2022 Games, Olympic bidders had to provide information on existing transport infrastructure (only roads and public transport), planned transport infrastructure independent of the Olympic Games venue, additional infrastructure independent of the Olympic Games venue, and additional infrastructure needed for the Olympic Games venue (Kassen-Noor, 2017). In addition, the time distances between venues, between Olympic villages and between the airport and the main training centres are considered other key elements to ensure an Olympic bid[54]. However, bid documents are often significantly modified and replaced during the Olympic process without respecting the original design[55]. Therefore, bids will always be shaped by the intentions of the Organising Committee and the central government of each host state. The host city contract includes several annexes, including the Technical Transport Manual, part of the city's contract with the IOC for compliance with Olympic rules.

The documents distinguish between nine transport planning and operations topics:

- Infrastructure and facilities.
- Customer services.
- Fleet operation.
- Bus operation.
- Public transport.
- Transport at the venues.
- Traffic management.
- Transport information.
- Planning and support services.

Finally, transporting athletes and essential personnel, such as judges (AT and TF), is highly time-sensitive and consequently very vulnerable. The IOC aims to provide them with safe and reliable systems that minimise their discomfort (Kassen-Noor, 2017).

In addition, the prerogative is that Olympic athletes must always have free traffic conditions between the Olympic Villages and the competition and training venues. Therefore, the leading Olympic bus network has become a standard feature for allocating a bid.

In 2004, Athens, for example, developed an Airport Express Facility, essentially a terminal used exclusively by members of the Olympic family (Odoni et al., 2009).

---

54   One of the evaluation criteria of the Olympic bid considers the distance between Olympic venues.
55   For example, Rio, after being announced as an Olympic city, expanded its proposed rapid transit system (BRT) from two to four lines.

2.3.3. *Olympic transport management*

As noted above, the management of Olympic transport is an essential element for the event's success. Hensher and Brewer (2002) identified five transport pressure points: trains, buses, taxis, airports, roads and parking. Based on these pressure points, Currie and Shalaby (2012) suggested several transport demand management measures grouped into six categories: travel capacity-building measures, travel behaviour change measures, traffic bans and emphasis on public transport (Kassen-Noor, 2017). Through constant real-time traffic and monitoring, intelligent transport systems have become the key to responding to temporary disruptions during the Olympic event. Since Sydney 2000, each Olympic city has provided two different transport systems. A primary network for athletes, members, and journalists, and a secondary network for visitors and participants in the management of the Olympic event. The primary network is exclusively for groups using priority lanes in the Olympic area. The secondary network ensures the connection between venues for Olympic ticket holders.

New or heavily revised traffic management centres, intelligent transport systems, surveillance cameras, and variable message signs are routinely implemented to redirect routine traffic, imposing new driving and parking restrictions around city centre venues and the Olympic ring (Kassen-Noor, 2013). However, the following measures have improved traffic flow, reducing delays for Olympic members and athletes. For this reason, planning an exceptional and priority transport service in the host city may compromise the viability of citizens in Olympic venues.

2.3.4. *The legacy of Olympic transport*

Olympic transport legacy involves physical transport infrastructures such as new or upgraded airports, roads (motorways, motorways, arterial roads, BRT routes), railways (metro, suburban, national, regional and suburban rail, tramway, light rail), marine (ferries), cable transport systems, new or upgraded off-site transport areas (car parks, loading zones, waiting and staging areas) and access to roads and loading zones, as well as transport facilities such as transport hubs, terminals and depots (Kassen-Noor, 2013). However, the legacy should be evaluated through public transport in the post-Olympic period.

Although many transport legacies are site-specific, Kassen-Noor 2013 identified six common legacies observed across the Olympic edition despite the different transport systems in place:

- New or improved connections between the airport and the city centre.
- Airport improvements.
- Creation and revitalisation of parks with high-capacity transport access.
- New high-capacity transport modes.
- Additional road capacity.
- Advanced intelligent transport system.

Therefore, the legacy of transport must be observed over the long term to assess the degree of acceptance by the population. One of the tools that host countries have frequently used as an infrastructural strategy is the provision of an intelligent transport system.

However, today, it is mandatory for candidate cities to integrate the planning, management, and use of Olympic transport systems into their projects strategically. Tokyo, Grenoble, Sapporo, Seoul, Barcelona, Nagano, Sydney, Turin, London and Rio were some editions that included transport in the infrastructural development of their territories. After London 2012, new global benchmarks were set for including transport planning in the strategic planning of Olympic cities. However, other editions, such as Sochi, Rio and PyeongChang, were in very different circumstances and required a rethinking of transport strategies for the event and the post-event phase. New regulations, the 2030 Agenda, and new Olympic bid measures are forcing host countries to find new modes of transport and cost-effective transport solutions to achieve sustainable development goals. Providing an intelligent and dynamic transport system will be one of the unique challenges for the bid cities (della Sala, 2022).

CHAPTER 3

The Territorial Impact of the Olympic Games

ABSTRACT

Considering the transformation of places, the different spatial representations of the Olympic territory will be analysed using the scheme created by Harvey (2002) and adapted by Dansero and Mela (2007) on the Olympic event. The next element will be applying the concept of territorialisation advanced by Turco in 1988 to analyse the production of the Olympic territory. Considering the local Project and thanks to the contributions of Magnaghi (2000) and Dematteis (2005), it will be possible to observe the different local development policies for host territories. Finally, the chapter will analyse the new concepts of Olympic legacy given the five dimensions the International Olympic Committee put forward.

3.1. The size of mega-events

3.1.1. The territorial impact of the Olympic Games on the host city

As noted above, the mega-event manifests through an extraordinary experience that implies reconsidering the ordinary politics of host venues. The society of spectacle and consumption in which we live is strongly linked to the different mechanisms of space production in Lefebvre's (1991) theories. However, the spectacularisation of the mega-event influences everyday life, the socio-cultural system, planning and architecture of places. The Olympics have developed through a capitalist culture that continues to be exploited by the world's largest metropolises to conceal other problems in the capitalist production system. Furthermore, capitalist influence allows the identities and qualities of places worldwide to become a homogenous and standardised space (Debord, 1991).

In this way, places seek to become a model of urban development based on the spectacle and the ephemeral. Undoubtedly, with the assignment of the mega-event, cities risk turning into theme parks, neglecting the local context and the problems of everyday life.

The theoretical debates on time and space within the social sciences are linked to the thought of Lefebvre, who states that space is socially produced. Lefebvre, like Foucault, links power and knowledge, asserting that the ruling class uses knowledge to maintain its temporal hegemony (Lefebvre, 1991b).

The Olympic event allows a community to reflect on itself, introducing permanent changes (Haugen, 2005).

Constructing a new society based on spectacle is a fragile and uncertain instrument of unification (Harvey, 2002). Therefore, the rediscovery of a national or local identity should be seen as a feeling of breaking social hierarchies. In conclusion, the impact of the mega-event on the host community should plan the best future scenarios to find new agendas considering the latest social phenomena, such as gentrification and segregation.

3.1.2. Analysis of mega-events

The analysis of the mega-event through the discipline of geography involves the consideration of two different approaches. A rationalist school and a humanist school. The rationalist school considers space as the primary locus for analysing the impact of the mega-event on the territory. The humanistic approach places the individual, amid the place, as the subject at the centre of the interests of the humanistic sphere, whereby the concept of place refers to ideas relating to the type of place and non-place (Auge, 1992), exploited by individuals[1]. For this reason, the rationalist view considers space as a rational interpretation of the subject, whereas in the humanist view, it is emotions and symbols that relate the place to the subject, consequently influencing the outcome and representation of space.

1   Auge, M. (1992). *The "NON-PLACES" spaces of anonymity an anthropology of overmodernity*. Edition de Seuil.

Over time, the two approaches have had a different level of application. Examples are zoning and spatial planning, tools that respect the rationalist vision. Participatory planning, on the other hand, is a humanistic practice that has yet to be used over time. The consideration of modern society as a liquid element is a metaphor for understanding how today's society suffers and is in constant and irretrievable change (Bauman, 1998). The liquid adapts and changes shape in its space, changing easily and making it impossible to stop it quickly.

For this reason, the following book aims to analyse the mega-event through two different approaches to studying territorial transformations related to Olympic urbanism in general and, specifically, to the Olympic Villages.

On the one hand, the territorialisation model proposed at a geographical level by Raffestin (1984) and Turco (1988), updated by Magnaghi (2000), will be adopted, while on the other, the territorial systems model proposed by Dematteis and Guarrasi (1995) will be used.

The first approach seeks to consider the territory's production at different levels, while the second is oriented towards the representation of local development processes from a territorial perspective.

The study aims to achieve a geographical reading of the Olympic event by analysing Olympic urbanism and its spatial dimensions over time. In this sense, it is a question of considering the event as an extraordinary production of the project territory and an ordinary production of the context territory.

3.1.3. *The social role of local transformation*

As we have observed in the previous chapters, the mega-event is a set of material and immaterial transformations and processes concentrated in a specific time and space. The mega-event affects the relations system in each place, producing new values and symbols that allow us to reinterpret the evolving reality. The theoretical consideration of the community will enable us to observe society through its everyday actions, the social actors and the ideologies that direct the behaviour and habits of today's society (della Sala, 2022).

Through the contribution of Dematteis (1995), we can consider two means of intervention in local communities: directly through inputs or indirectly stimulating local self-organisation processes.

Thus, the processes manifest themselves in two different spatial dynamics (of places, cities or regions):

I. As a simple valorisation of space: a mechanism by which certain given local conditions and resources are transformed into advantages compared to changes in global economic and political relations (variations in demand, prices, alliance, wars).

II. Local development is a process in which certain comparative advantages are more or less the consequences of local actors' self-organisation.

In the first case, places and territories are seen as passive support for global processes. This can have consequences in terms of devaluing the local community.

In contrast, the development of local territorial systems plays an active role in producing and creating regional synergies[2].

Moreover, synergies can initiate new self-cumulative local processes based on cooperation and competition. Thus, local actions should respect the community's requirements and represent each specific place's possibilities regarding its natural resources. Considering Olympic sites as nodes entering a global network makes it possible to observe the connections and interconnections that can put the host community at risk. An international project will have no condition for the existence and validity of local practices. Moreover, as observed in Barcelona, Sydney, and London, a global scale will permanently transform the regional scale. Participation and exchange of information between the community and the organisers can help to improve post-Olympic proposals and expectations without

2   In this way, the spatial Project does not represent objects but subjects, placing itself at the centre of the territorial development project (Dematteis, 1995).

compromising the organisation of a global event. Establishing dynamic and interactive connections between the community and all *stakeholders* will allow for developing new participatory practices that consider the local scale as a measure for organising a global event.

3.1.4. The spatial representation of the territory

The production and organisation of the Olympic space are strongly conditioned by the spatial representation; therefore, the following elements are constituted as results of the interventions. The representation of the Olympic territory is the reality that the organisers want to transmit to the world and represent the whole world. The Olympic event can introduce new areas and elements of territorial representation of different scales.

For example, spatial representations of the Olympic territory sometimes hide information and reduce the distance between Olympic venues[3]. On some maps, distances between competition venues must be specified or include places of cultural interest in the territory. The following lack of information allows tourists and participants to observe a temporary territorial space without consideration of the elements represented in the permanent territory, disadvantaging the experience of the local community.

Lefebvre (1991) identifies three dimensions for the observation of spatial production:

- Material spatial practices: defined as the experiential dimension of space;
- Representations of space: symbols, meanings and codes make it possible to understand, through the perceptual sphere, the practices of spatial production;

---

3   On the Milan-Cortina 2026 Organising Committee website, the Olympic map omits the city of Turin. The exclusion of Turin can be seen through a hypothesis of competition between the two cities.

- The spaces of representation: the reproduction of more complex symbols, discourses and utopian programmes belonging to the collective imagination sphere.

Subsequently, Harvey (2002) argues that Lefebvre's three dimensions must be linked to four fundamental attributes:

- Accessibility and distance;
- The appropriation of space;
- The naming of the space;
- The production of space.

Therefore, the scheme advanced by Harvey (2002) and readapted by Dansero and Mela (2007) about the Olympic event effectively lends itself to categorising the different spatial dimensions driven by a mega-event. The column relating to the material space is composed of the transformations permanently established in the territory through the physical and socio-economic components. Meanwhile, in the column on the representation of space, an attempt is made to examine the elements included in Harvey's three dimensions. It has been shown that the next dimension is always represented in all documentation and evaluations of the Olympic programme. Finally, the column relating to spaces of representation indicates the various forms of subjectivity linked to the Olympics concerning possible destinations and development models. However, the central aspect is represented by the effects imprinted on the collective imaginary and guiding the actions of individual and collective subjects (Dansero & Mela, 2007).

In the previous sections, we have analysed the territorial space up to the definition of a local territorial system. Before exploring the concept of local communities and territorial organisation systems, going deeper into the idea of territory is necessary. To analyse the territory, we will use the territorialisation model given by Raffestin (1981) and Turco (1988) and applied to mega-events by (Dansero & Mela, 2007). We will observe how territory production through acts of territorialisation can determine different outcomes and strategies in the post-event phase.

Table 13. An application of Harvey's scheme to an Olympic event (Source: Translation from Dansero & Mela, 2007, p. 12)

	Material space	Representation of space	Spaces of representation
Absolute space	- The Olympic territory "project" (venues, sports and host facilities, infrastructures) - The "context" of the Olympic territory, with its physical, geographical, demographic, etc. characteristics. - The territorial jurisdiction of the bodies and persons involved (IOC, national committees, local authorities, etc.). - Territorial scope of the various heritage enhancement projects	-Cartographic depictions of the Olympic territory - Maps of places, tourism opportunities, etc. - Thematic maps depicting the distribution of specific phenomena in the Olympic territory -Objective" descriptions of the landscape - Maps intended to control the territory	-Feelings of inclusion/ exclusion concerning the project territory and the context territory. -Role feelings related to the Project's and context's territory. -Sense of security/ insecurity in space - Feelings of control, of power
Relative space	Friction role of distance (in terms of cost, time, etc.) in relation to the presence of flows: - financial, - of people (tourists, sportsmen and women, accompanying persons, economic operators), - information (to organise the event, evaluate its effects, and plan new initiatives). -of the images transmitted by the media to and from the Olympic territory - flows of matter and energy that define the balance of the ecosystem in the Olympic territory	Non-Euclidean metric representations of the project and context spaces -Representations related to various forms of flow: e.g., mobility of Olympic players and spectators, energy and water consumption, etc. -Functional representations for flow control	- Feelings related to the nodal role of spaces concerning international flows -Feelings of inclusion/ exclusion about the flows generated by the event -FFeelings of familiarity/ restriction concerning a condition of high flow intensity

(continued)

Table 13. Continued

	Material space	Representation of space	Spaces of representation
Relational space	-Relationships with the social, territorial and environmental resources of the area. -Internal social relations (system of complementarity and/ or competitive and conflicting relations between the parties in the area) - The relationship between the Olympic Area and the broader regional, national and European context. Socio-economic and cultural sedimentation processes (social capital, institutional capital, etc.).	- Representations of value, referring to facts and subjects in a space of different nature (economic, cultural, social, etc.). -Representation of the landscape as a set of values attributed to the territory. - Representations and icons of Olympic values (logos, uniforms, mascots, etc.)	-Feelings of cultural proximity/distance to the Olympic spirit and the atmosphere of the Games - Feelings associated with the new symbolism of places brought about by the Olympic experience (before, during and after the games) - Visions of a possible future project linked to the enhancement of the Olympic heritage

3.2. Olympic territorialisation

> An urban structure entirely generated by the laws of economic growth; with a strongly dissipative and entropic character; without boundaries or limits to growth; unbalancing and strongly hierarchical; homogenising the territory it occupies; eco-catastrophic; devaluing the individual qualities of places; lacking in aesthetic quality. (Magnaghi, 2001)

The territory is defined as a historical product and artefact resulting from a long-term evolution between human settlement and the environment. Therefore, the territory is considered a living and highly complex organism. An ecosystem in continuous transformation is the product of the encounters between culture and nature and is formed by places endowed with identity, history, and character, including territorial and urban "typologies" (Magnaghi, 2001).

Table 14. Spatial practices and mega-events (Source: Translation from De Leonardis, 2006, p. 51)

	Accessibility and distance	Appropriation and use of space	Dominance and control of space	Spatial production
Material spatial practices (experience)	Increased flows of money and people. Increase in immigrant labour for the production of new spaces. Improving information technology facilities to achieve excellence. Changes in transport infrastructure (roads, trains, metro).	Increased construction. New tendering facilities. Privileged transport and communication networks. Increasing polarization and gentrification phenomena. Rearticulation of productive, recreational and residential spaces within the urban territory. Use of space by new social actors (volunteers).	Exclusive control of public areas by the organisers. Private management of public spaces. Controlled and protected spaces against antagonistic movements and communities considered socially dangerous. Exclusive spaces are reserved for the protagonists of the event (villages).	Production of new urban and intra-regional connections. Redistribution of urban centres and passenger flows. New architecture and urban symbols with international appeal.
Representation of space (perception)	Increasing accessibility in spatial terms. Development of new spatial maps. New spatial communication through signage.	Change of mental maps in terms of distance and perceived space. New centralities and hierarchies at the regional level. New meanings and uses are attributed to places already steeped in history and identity.	Production of spaces to enhance the strategies of regional communities in a more open and competitive economic and social scenario. Local pride. Building consensus and monitoring potential threats.	New urban and tourism marketing strategies. Re-imagining the image of places through guides, photos and locations.

(continued)

Table 14. Continued

	Accessibility and distance	Appropriation and use of space	Dominance and control of space	Spatial production
Spaces of representation (imagination)	Instrumental use of the media. Appropriation of spaces for commercial and artistic purposes.	Global popular participation. Familiarity of places experienced in the media. Popular appropriation of public spaces.	Rediscovery for the new tourism market of new routes and new products. Invention of social geographies. Mutual enrichment and contamination.	New meanings of territorial identity.

The first decade of the twentieth century, dominated by Fordism and mass production, made the organisation of the territory more complex about the economic growth of places. Therefore, major events such as the Olympics or Universal Expositions imply the use of different territories that must coincide with the economic growth of the places[4]. As noted above, to analyse Mega-events through a territorial analysis, we must look at the territory through two different geographical representations in terms of form, control, governance, symbols and values. The first representation refers to the "project territory", represented by the Olympic territory where the structures and functions necessary for the event's organisation will be built. The second representation refers to the "context territory", considered as the existing part of the territory observed through different scales (regional, national, local)[5]. We can observe a potentially conflictive local-global relationship at both levels at work (Dansero, 2002). However, the conflict will remain until the closing ceremony of the Paralympic edition. Therefore, the definition and construction of the Olympic territory, being a spatial-temporal system, implies a territorial organisation and a local effort that can be considered a *stress test* for the whole community. As we will observe in the next chapter, the spatial structure of the Olympic Games continues to change in a multipolar system, becoming more complex and more extensive every day. During the Olympic event, the territory will have to provide new memorable lines of communication for the transfer of the plurality of actors (athletes, technicians, media, public, organisers, volunteers, sponsors)[6] for Olympic transport systems.

As noted in the previous section, transport management has become an essential element for the event's success throughout the editions of the

---

4   The identification of the Olympic territories imposes the analysis of different territories in a geographical space, which in the Beijing 2022 edition will reach up to 200 km from the Olympic city.

5   The boundaries of the project territory should be considered temporarily only for the exploitation of the event, while the boundaries of the context territory during the Olympic event may compromise territorial control and governance.

6   The contribution of Professor Eva Kassens-Noor (Kassens-Noor, 2017) is recommended.

modern Olympic Games. Moreover, the organisation of a temporary transport system modifies the shape of local systems, causing new journeys and congestion of the traditional transport system. As we have observed in the previous section, the maps of Olympic events draw a territory different from the geographical representations of the host territory. During the event, temporary operators and actors govern and control the Olympic territory. Moreover, some areas of the Olympic territory will be inaccessible to most citizens[7]. The following complex and temporary spatial structures are contained in the host community's territory, which has its logic and operating regulations.

Analysing the host territory during the mega-event planning, it can be observed how the Olympic event implies constructing a spatial structure promoted at an international level, which highlights or excludes some localities rather than others. Consequently, the Olympic spatial structure presupposes a transformation of the local scale to adapt the territory to its temporal needs. However, the construction of an Olympic space presents a homologate and standardised theory with its own rules for producing the Olympic territory, shaped by the experience based on previous editions.

The big event simultaneously seeks and consumes spatial differences but may produce them (Dansero & Mela, 2007). The outcome depends exclusively on local and national administrations, which must negotiate and mediate between the homologate tendencies of the IOC and supralocal actors. As we will observe in the case of Turin 2006, in Chapter 6, it is difficult for local communities and organisers to implement strategies that do not result from conflicting national and global visions. Thus, transforming the territory during the mega-event implies complex management during and after the event. However, the results will only be observed in the post-Olympic period.

Parallel to the Olympic territory, there is the "context territory", made up of the Olympic city, the peripheral localities and all the territories "crossed" marginally by the big event. The context territory undergoes changes that

---

7   For example, the Olympic Village and the International Venue are areas with rigorous and limited access for most Olympic operators.

permanently affect the territory, redefining the local scenario and the hierarchies of the localities in the regional framework.

Contrary to the Olympic territory, the context territory is controlled, organised and governed by fully public management composed of all the local authorities and the private capital of the target territory. The relationship between the two territories demands and requires in-depth work in realising the different structures that will be added to the urban and territorial images of the Olympic sites. In this context, participatory planning can help define strategies for constructing a new collective image. Thanks to the coordination and contribution of all local actors, the regional scale benefits without compromising the promotion of the territory on a global scale.

3.2.1. Territorialisation

> We can imagine territorialisation in a strictly chronological perspective, as sequences of acts that begin in a distant prehistory. (Turco, 1988)

As we observed in the nineteenth century, the agrarian and industrial revolutions represent the fundamental points of eliminating the limits of action and, in short, of the stages in constructing an artificial environment. Furthermore, given the previous paragraph, territory is seen as the result of applying labour to a defined space. Thus, territory can be seen as an extension endowed with certain properties over which human action is exercised.

The acts of territorialisation[8] Represent the whole of a territorial mass in space. The morphology of the territory of a given section of the earth's surface is subject to various exogenous and endogenous transformations depending on various natural factors. Earthquakes, wars, etc., territorialisation is thus a significant process by which space incorporates anthropological value (Turco, 1988).

---

8   Considering Raffestin's (1981) and Turco's (1988) contributions, territorialisation is regarded as the production of territory in a space produced by the actions of territorial actors.

In this way, the process of territorialisation must be seen as continuous growth in a territory that continues to reconfigure and readapt itself to the demands and habits of the communities.

The analysis of territorial acts must start from three fundamental categories (Turco, 1988): naming, reification and structuring. Consequently, Olympic territorialisation, conceived as the production of new temporal territory, is intertwined with ordinary transformative dynamics through a T-R-D cycle that can be read as the encounter-contrast between different territorialising acts at work and can be traced in three categories (Turco, 1988):

- Designation
- Reification
- Structuring

Applying the territorialisation process in a significant event starts from the bidding and site selection phase, passing through the de-territorialisation phase that follows the event and coincides with dismantling the Olympic territory. The re-territorialisation phase is strongly linked to the abandonment and the Olympic infrastructures that need to be reconfigured and used over time. The last phase manifests itself only in the post-Olympic phase and continues to transform the heritage and territorial capital permanently.

For this reason, one can hypothesise an Olympic territorialisation – de-territorialisation – de-territorialisation (T-D-R) cycle that generates conflicts with the T-D-R processes of the contextual territory. According to Turco (1988), the act of naming places is a social work that is accomplished through cognitive and communicative strategies. The attribution of a specific name to a temporary place allows it to enter the global and spiritual sphere of communities. Meanwhile, the second form of control manifests itself through reification, which consists of the occupation of space by using its resources. Symbolisation is followed by the material control of acts of territorialisation. Successively, the last form of control is structural. Thus, structuration is regulated and organised through a process of control by

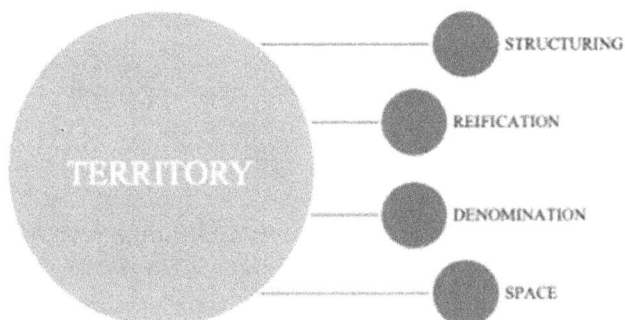

Figure 5. Space and territory: Acts of territorialisation (Source: Own elaboration on Turco, 1988, p. 78)

authorities who, through the use of norms and rules, manage the territory about the objectives of the specific territorial context (della Sala, 2023).

3.2.2. Production in the territory: The denomination

The preceding section established that the territorialisation of mega-events is strongly related to the territorial symbols associated with the event.

For example, in the Olympic Games, we can find the Olympic ring, the Olympic Village, the Olympic Stadium, the Olympic Square, the Olympic trampoline, etc. ....

Therefore, the denomination can be transitory, existing only during the games, or survive with the territory, becoming a pretext for constructing new strategies for different material and immaterial purposes[9]. For this reason, the promotion of places is directly related to the exploitation of the event in the medium and long term (della Sala, 2022). For example, in Barcelona (1992), the Olympic Village and the Olympic Port were part of a territorial marketing strategy that continues to mark the importance

9   We have observed different models of preserving the Olympic name throughout
     Olympic history. For example, Barcelona has created an Olympic Quarter, a port
     and a metro line. London and Beijing have turned the Olympic space into a large
     sports park, which changed its name after the Olympic event.

of the Olympic heritage in the territory. The naming of places is thus the first act of territorial production.

Consequently, naming the Olympic Games begins with the bidding process, defining the event's territory and associating the event date with the host city (Barcelona 1992, Sydney 2000, Torino 2006, London 2012, Rio 2016, etc.). A temporary and symbolic designation that becomes a permanent fact over time through a process of legitimisation, which tries to attract new global images of the Olympic territory. The permanent legitimisation of a designation involves the construction of an identity, which is transmitted globally over a long-term period. In the same way, a symbolic control of the territory complexifies the results and the legacy of the meme mega-events of the territory[10]. Regarding the Olympic Games, it is essential to remember that only the host cities can use Olympic symbols to promote the territory during the bidding phase and the event's organisation. However, the naming of Olympic sites involves a renewal of the images of the territory or the landscape, through technological tools such as television, the internet, and photographs, allowing the city to be promoted on a global level. Therefore, the danger is that only some Olympic territories and works will be affected by the permanent naming of the sites. For example, the territories crossed by the temporary transformations of the Olympic space will disappear after the closing ceremony of the Paralympic event. For this reason, the naming feeds new communication strategies for the exploitation of the tourism market as a consequence of mega-event[11].

Many criticisms have been raised over time regarding the temporary laws that venues must adopt to sell and market advertising space on Olympic territory.

> Significantly, the municipality and the IOC urged the international media to use the name "Torino" instead of the international Turin. (Dansero & De Leonardis, 2006)

10   The temporary production of information can lead to the disappearance of the Olympic effects in the post-Olympic period.

11   It is essential to affirm that, during the Olympic event, the Olympic territory is protected under the control of the IOC concerning the space for the use of billboards in the whole Olympic area.

3.2.3. Production in the territory: Reification

The second process, reification, is considered the material transformation of the territory and is the most evident and lasting aspect of territorialisation. Thus, the construction of the Olympic works, the infrastructure, the sports facilities directly related to the event and all the works connected to the event's organisation constitute the material transformations of the contextual territory. It is essential to note that reification can catalyse other artefacts over time. Structures and infrastructures are formed as the material heritage of places, implying new future transformations for the use of spaces (della Sala, 2023). For this reason, the Olympic reification is constituted by an extraordinary material transformation with the help of large international architectural firms that can promote themselves by performing Olympic work. On the other hand, material transformations can be accompanied by specific technological innovations that, in the future, will achieve the status of catalysts for new infrastructure construction and organisation processes.

For example, Sydney 2000 proposed the construction of a sustainable Olympic Village through the use of renewable energy sources. Turin 2006, on the other hand, proposed the construction of the Olympic Village using recycled materials. In conclusion, Olympic reification is a complex phase that may generate different outcomes in the post-Olympic period.

3.2.4. Production in the territory: Structuring

Structuring, as we have seen, represents the last phase of the territorialisation cycle, a process that introduces sensory or structural control, selecting areas and encouraging the construction of territorial structures that can guarantee the implementation of programmes and strategies (Turco, 1988).

This phase can manifest itself in mega-events in the post-Olympic period. Two possible levels can be identified in this regard:

- Sub-regional level (Olympic map)
- Local-level (Olympic venues)

However, the places in the Olympic spatial system will be affected over time, becoming *glocal* places.

For this reason, during the event, the Olympic territory may be affected by the integration of multiple institutional bodies acting with different strategies and pursuing different objectives. In this sense, regional and strategic plans are considered fundamental elements of organising and defining common objectives representing all the actors involved in the Olympic Project. Strategic spatial planning reduces the risks of conflicts in spatial systems, and a functioning system can be imposed for the post-event phase. On the other hand, the structuring of the territory may reactivate new economic, social and urban processes that will last over time (della Sala, 2023). The territorial structure promoted by the big event can take on different forms and impacts depending on the objectives and activities indicated by the territorial actors about the post-Olympic legacy (Dansero, 2010).

The territorialisation phase is followed by the de-territorialisation phase of the spatial-temporal system developed for the Olympic event. The de-territorialisation phase may involve a re-territorialisation phase where the transformations of the context territory are resolved or increased. Over time, we have observed different options for the post-event transformation of Olympic sites at the Olympic Games:

- Dismantling,
- Reuse,
- Reconversion,
- Abandonment[12].

Throughout Olympic history, we have observed different projects that were often overestimated and, therefore, have produced an excess of territorialisation in the context of territory, remaining as artefacts without

12   Abandonment is seen as defeating the territorial Project and, therefore, destroying the contextual territory.

identity. For this reason, the potential estimation risk may affect many communities (della Sala, 2023). Lack of experience and resources may introduce a conflict between the city and the Olympic sites (Jennings, 2012). For example, in Turin 2006, as observed in the doctoral study (see della Sala, 2022), the ski jumping facilities and bobsleigh track left a fracture in the Piedmont region that can never be recomposed. Moreover, these works can be a source of conflict between the regional territory and the central government. Consequently, constructing overestimated facilities, structures and infrastructures to host mega-events expresses central governments' objectives and will to use mega-events to promote other symbols, monuments or images (Poynter, 2016).

It also needs to be said that the IOC has different interests from the host states. The form of Olympic territorialisation is often totally different from the initial plans of the Organising Committees. Only by reconfiguring the rules and the Olympic contracts for the bidding cities will it be possible to reduce the territorialisation processes that are not necessary for the host communities in the post-Olympic period (della Sala, 2023).

As we have observed in the preceding sections, illustrating some theories and possibilities induced through mega-events, new possibilities have been identified for creating new territorial systems that can be transformed into tangible resources to exploit the big event at the territorial level. To investigate the development of local systems, it is helpful to introduce Magnaghi's contribution (Magnaghi, 2000) to the local Project and its development perspectives. Subsequently, a brief introduction to the SloT theoretical model (Dematteis, 2005) will be made, trying to identify the different development perspectives that can be manifested through the mega-event. Finally, some considerations will be advanced on the possibilities for local communities to take advantage of mega-events heritage.

3.2.5. *The local project: Local development policies*

The local project is the political manifestation of a demand, a need, an idea to respond to the challenge of globalization. (Magnaghi, 2000)

Table 15. The cycle of territorialisation in mega-events (Source: Own elaboration on Turco, 1988)

	Territorialisation	De-territorialisation	Re-territorialisation
Designation	• Venue of the event • Territory of the event • Facilities • Structures • Infrastructure • Squares, Monuments		• Places, structures, and artefacts that still retain the adjective "Olympic". • Creation of a territorial brand • International events market
Reification	• Sports facilities • Accommodation/ Hotels • Olympic Village • Olympic structures • Infrastructure	• Dismantling • Abandonment	• Public facilities • Multifunctional facilities • University Campus • Popular residences • Hotels • Infrastructure • Urban spaces, theme parks • Tourist accommodation
Structuring	• Regional territory • National territory • Local Territory • Glocal territory	• Reuse • Reconversion • Liquidation of the organising committee	• Reorganisation of territorial promotion bodies • New spaces for international audiences • New post-event organisational structure • Setting up a volunteer event organiser • Creation of a specific structure for mega-events

In the twenty-first century, the local Project is seen as the possibility for local communities to resist the exclusivity of the metropolis, defending their identities, traditions, cultures and landscapes without compromising social relations and a sense of community. On the other hand, the local systems of Olympic cities in the post-Olympic period will conflict with the central city's objectives and will be subject to the different rules of the global mega-event market (della Sala, 2022).

Furthermore, after the Second World War and the end of post-Fordism, the territory has become a place of value production. Therefore, value should not only be considered from an economic point of view. Awareness and symbols enable community recognition and increase the value of territorial heritage in creating lasting, long-term wealth.

The construction of the local Project is based on the pact of a plurality of actors (Magnaghi, 2000). The definition of the local Project is a complex process of clarifying conflicts, defining objectives and redefining projects to implement a shared project that can generate heritage value for the community.

However, the construction of the local Project should be funded and supported by a statute of places, seeking to overcome the collective individualities of places and propose a common goal through new forms of participation and direct democracy. Through the constitution of a shared statute, local communities will support territorial transformations through rules, norms and pacts, supported by the community.

This way, the local Project can foster and promote self-employment, crafts, cultural districts and micro-enterprises.

"The local project presupposes the growth of the powers and competencies of municipalities and supra-municipal territorial entities, expressions of the municipality as a higher local authority" (Magnaghi, 2000). The construction and promotion of internal democratic institutions (local development agencies, pacts, dialogue tables, participation workshops, living labs) can provide a solid basis for the promotion of local policies and networks at the regional level. For this reason, governance and the different levels of control of internal democratic institutions are current and crucial issues for the sustainable development of our local communities.

"In the glocalist hypothesis, local development takes shape to the extent that the local community is contaminated by the global, bringing to the local the innovations coming from the opening of relations between long and short networks; local development occurs when local society can build horizontal networks in the global system" (Magnaghi, 2000). Network forms are combined in-depth in existing local areas without the locals being able to exit the global Project. The intersection of new global networks for local communities and hosting mega-events is one of the most significant risks to the preservation of the authenticity and autonomy of local communities. Consequently, the local territories of mega-events are embedded in a global network of local societies, strengthening themselves through new relationships and networks that contrast with current centralist forms of economic globalisation.

Local communities can establish:

I. Inter-local information relations and solidarity networks that interconnect with global networks.
II. The proliferation of cities capable of building non-hierarchical global relationships through the diffusion of services in peripheral regional networks, in response to concentration processes.
III. Eco-solidarity business and financial relationships that develop local networks and transfer to the global market.
IV. Self-sustainable local production systems based on the valorisation of heritage.
V. Networks of local development agencies that interconnect top-down projects with bottom-up projects.
VI. South-south, south-north cultural relations that densify the overlapping wefts of north-south networks: self-representations versus representations of the centre (Magnaghi, 2000).

Therefore, in the presence of a mega-event, the global effect is overflowing; it should try to preserve local communities, activating different policies, actions, processes and projects that allow for:

- Strengthen the internal relations of each territorial system by constructing new social fabrics to express each territory's peculiarities and capacities within a framework of sustainable development.
- Develop and build networks between the local and the supra-local in the medium and long term. The new intra-local networks should modify the hierarchical system of metropolises and global cities towards a complexification and multiplication of regional systems.

Consequently, by strengthening relations and developing new intra-local networks, a new intra-local system will be created. A system that will favour the construction of new relationships in a way that catalyses new processes of solidarity exchange and participation. Through mutual respect for global economic networks, new eco-flows will be established in the Olympic cities in the post-Olympic period (della Sala, 2022).

3.2.6. *Local, territorial system*

The model of local territorial systems (SLoT) advanced by Dematteis (2005) allows us to use the model to describe possible social interaction relations for territory, governance and sustainable development. By introducing the SLoT theoretical model, we will look at the different perspectives of local development that can emerge in Olympic territories during the post-Olympic period. The hope is that it offers a contribution that allows us to reflect on the construction of a local, territorial system that is not affected by the global context of the mega-event perceptual model of territorial systems advanced by Dematteis (2005) is composed of the following elements:

1. *The local network* is constituted by all the relationships and interactions between all the existing subjects that can be developed in a local area to construct common objectives of local development. The term local identifies the Project's geographical scale, allowing interactions about the physical proximity of the places and

communities that comprise the local territory. Communication, exchange, knowledge, history, heritage and standard practices of the local territories allow the development of medium- and long-term relationships on the target territory. The SLoT model can be identified and developed through local subjects, which advance a collective development project without compromising the autonomous development of local particularities. The transformation, reorganisation, requalification and development of the territory should be based on a shared vision of the local territory to exploit the synergies of the places.

2. *The local milieu*: identifies a set of permanent places through socio-cultural and morphological characteristics identified in a given geographical area through the heritage and history of each place. The territorial capital of each place is constituted through a set of processes, objectives and resources that favour the subjective possibilities of each place. The representations of the territorial space will be the manifestation of each local particularity which, thanks to all its activities, can improve the environment of the communities and transform the territory based on the demands of the local communities.

3. *Local network relations*: it constitutes the whole system of local ecosystems in a local network through the shared values and objectives provided by the main actors of the territory. The process of transformation and redefinition of local networks is considered an intangible element for the exploitation of common values over time.

4. *Interactive relations between local and regional networks*: these are identified in the communication processes between the local and supra-local levels (regional, national, European, and global). Relations between different levels of intervention constitute a dialogue table, where relations with the local environment must attract new exogenous values from mega-events without compromising local interactions and the evolution of a sustainable territorial development network over time.

The SLoT model, as we have observed, is constituted by the identity of the local communities and the organisation and hierarchy of the system in the medium and long term (Governa, 1999). For this reason, the organisation of the local system in mega-events must inevitably consider the different particularities of the territory and the different local development policies that the supra-local level must know and build at the national level (Poynter, 2010). As we will observe in the case of Turin 2006, the organisation of governance at the regional level was a fundamental element for the exploitation of the Olympic Project throughout the context territory. However, local development strategies should be implemented through shared planning and organisation over time in the post-Olympic period. In this sense, the mega-event is a stimulus for implementing and creating new strategies, which can constitute a model of local development for the future of the Olympic territories in the post-Olympic period.

Only through a project shared by the local community can positive results be obtained in the image and perception of the communities involved in the Olympic territory. The SLoT model can offer a series of processes and objectives, which must be continuously stimulated to establish a territorial system that bases its development on sustainability, participation and social cohesion (della Sala, 2022).

The concepts of Olympic heritage and legacy are strongly linked to the strategic planning of the territory in consideration of the host communities' peculiarities. However, dismantling, redeveloping, reusing and abandoning Olympic sites can permanently compromise the territory and citizens' perception of the organisers and the local governance system. The liquidation of the Organising Committee and the temporary structures needed to carry out the tasks related to the execution of the works and activities of the mega-event pose a significant risk to the local public agenda, as they may result in a vacuum and a decrease in popular support. Meanwhile, constructing one or more specific structures for implementing the new post-Olympic goals is the only way to reduce the chances of abandonment and failure of sustainable territorial development over time.

However, in recent years, follow-up activities and monitoring of post-event results have given increasing importance to observing the UN 2030 Agenda's Sustainable Development Goals. In the following sections, we will examine the Olympic impact assessment through the technical documentation of the IOC, which is recognised as the body responsible for managing Olympic works in Olympic cities.

3.3. The Olympic legacy

3.3.1. Introduction to the Olympic legacy

As noted in the previous sections, the relationship between mega-events and the tangible and intangible values of the event is complex and challenging to analyse. However, it should be studied over a medium to long-term period so that no promotional manipulation of the event is carried out on the territory. For this reason, the development prospects of host cities in the post-Olympic period are a topic of growing interest that can be defined as "Olympic legacy". The term "legacy" can be defined as the set of Olympic works, infrastructure, housing, projects and the Olympic experience. In the meantime, the concept of *legacy* is a new concept rooted in the philosophy of transforming host cities to encourage the spread of sport. However, international debate has focused on the difference between impact and heritage over time[13]. As we will observe in the next chapter, the principles and philosophies of the Olympic movement, founded by Baron de Coubertin, aimed to provide the territory with sports facilities to promote sport in the post-Olympic period. Therefore, the practice of sport is understood as a means of education with a catalysing power[14]. Of values and principles of respect, brotherhood and

---

13 Definition of Olympic legacy: Cashman (2005): (1) sport; (2) economy; (3) infrastructure; (4) information and education; (5) public life, politics and culture; (6) symbols, memory and history; Chappelet (2006): (1) sports legacy; (2) economic legacy; (3) infrastructure legacy; (4) urban legacy; (5) social legacy.

14 In Olympic history, Berlin, Oslo, Rome, Tokyo, Mexico City, Barcelona, Sydney, Turin, Beijing, and London are considered the most significant cities that have used the Olympic Games as a catalyst for sport in their territory.

overcoming limits. For this reason, the effects of the Olympic legacy must be observed from a multidisciplinary perspective, and some of its impact can be a change in image, economy, tourism and sports practice. Furthermore, the legacy can be found in different areas such as culture, economy, architecture, environment and territory[15].

"In addition to being determined by Olympic history, the legacy must be contextualised in different local contexts" (Dansero & Mela, 2004).

Other authors, such as Hiller, for example, propose replacing the term "legacy" with "outcomes" (Hiller, 1984). Hiller's concept introduces a reading of outcomes regarding sporting/non-sporting and programmable/non-programmable.

On the other hand, Cashman (2002), about the Olympic legacy, identifies some fundamental characteristics to understand the term *legacy*.

- Legacy can be tangible (hard) or intangible (soft).
- Legacy is not a monopoly
- There are different legacies for each actor involved in the event
- Other legacies are not linked to the celebration of the event

Cashman (2002) states that legacy includes both expected and unexpected aspects, which should not essentially be connected to celebrating the Olympic event.

The issue of the Olympic legacy grew from Atlanta 1996, where it was introduced in the post-Olympic reports. Subsequently, Sydney 2000 was the first edition to raise some critical points about legacy and post-Olympic planning in a cross-cutting manner. The Sydney project introduced specific

---

15   Measuring event legacies: Event tourism: Getz (1989, 1991), Hall (1992), Kang and Perdue (1994), Carvalhedo (2003), Dwyer et al. (2004), Chalip and McGuirty (2004) and Solberg and Preuss (2006).
        Employment impact: Ritchie (1984, 1996), Burns et al. (1986), Mules and Faulkner (1996), Hotchkiss et al. (2001), Hagn and Maennig (2007); Urban development: Evans (1995), Hughes (1993) and Meyer-Künzel (2001); Environment: May, 1995.
        Social impacts: Shultis et al. (1994), Hodges and Hall (1996), Lenskyj (2002), Fredline et al. (2003), Misener and Mason (2006) and Smith and Fox (2007).

sections on the Olympic legacy, tourism, economy, technology, sports facilities and social identity (Chalip, 2010). Since Sydney, the concept of
legacy has been introduced into official IOC documents and, over time,
has been incorporated into the *host city contract* and post-Olympic documentation. Today, post-Olympic legacy planning is essential for hosting
the Olympic Games.

Cashman (2002, 2005) introduced six categories to classify Olympic
legacy:

- Economic legacy.
- Legacy of the physical and built environment.
- Informative and educational legacy.
- Legacy of public life, politics and culture.
- Sporting legacy.
- Legacy of symbols, memory and history.

According to Cashman (2002), legacy should:

- Guaranteeing a city's return on investment.
- Fulfilment of tender promises.
- How to resolve outstanding issues.
- A return for the community.
- Avoid negative publicity and urban decay.

In addition, we find other equally important reasons why legacy is vital:

- Adopting a professional attitude is to be replicated in the post-
  Games period.
- Legacy is linked to the third pillar of Olympism, the environment.
- The legacy is linked to the balance of financing.
- The Olympic movement gains cultural capital through the places
  and symbols associated with the Olympic Games.
- Legacy can be perceived as a burden and can compromise planning
  outcomes over the years of preparation.

The international debate after Turin (2006) has evolved to the point where legacy has become the sponsoring element of the Olympic bid. Moreover, post-event legacy planning is now pivotal for future planning within the host territory. Furthermore, the definition of uses and the allocation of managing bodies of Olympic facilities is a fundamental process for the long-term exploitation of the territory. However, a lack of foresight can become misleading publicity for citizens. Unfortunately, we have witnessed overrated Sochi, Rio, and PyeongChang constructions, which must be addressed and decayed. The event's preparation, management and planning must be apparent from the outset so there is no criticism in the post-Olympic period. Therefore, a specific body should be set up to manage the post-Olympic legacy, just as London did in 2012. Since the organising committee has a limited duration, only by establishing a permanent structure can the risks of abandonment be reduced. Montreal, Athens, Turin, Sochi, Rio and PyeongChang are some of the experiences that have produced a significant post-Olympic management deficit. One of the risks for local communities hosting winter editions is that of creating *white elephants* (Cashman, 2002). However, one aspect that still needs to be studied concerns the temporal dimension of the Olympic legacy. International references do not refer to a specific period for longitudinal studies.

Moreover, the spatial and temporal dimensions must be explicitly analysed for each scale because these are different scales. Therefore, we can introduce a global scale, a national scale, a regional scale and a local scale. However, each of these scales needs continuous investment to achieve the goals for post-Olympic planning.

The IOC in 2010 suggested five dimensions for legacy research[16]:

1. The degree of planned/unplanned structure,
2. The degree of positive/negative structure,
3. The degree of tangible/intangible structure,
4. The duration and timing of a change of structure,
5. The space is affected by the change of structure.

16   IOC, Legacy and Impact, 2010.

Therefore, structure, space, and time are the factors to be considered when investigating the temporal dimension of the Olympic legacy[17]. The continuous investment and implementation of the plan allow cities to dynamise the objectives of the actual demands of the citizens[18]. However, the objectives of each specific context must respect the promises made during the bidding process. Moreover, it would be interesting to introduce a contract between the city and the citizens, as they are the main actors of the Olympic legacy. Bid reports often have high expectations, but the organising countries often leave them as significant unfulfilled projects. The non-fulfilment of certain expectations in Olympic cities feeds collective dissatisfaction, facilitating the creation of new movements against the organisation of mega-events.

The only solution is often adopting a long-term vision that includes the local community in the post-Olympic Project. An extended timeframe can reduce criticism from citizens. However, the Olympics can accelerate ordinary transformation processes, while on the other hand, it can destroy the image of the city and the public administration around the world. Finally, it is suggested that the post-Olympic legacy be planned so that the objectives are achieved and the processes can be applied daily as an example of good practice in local government[19].

3.3.2. The Olympic legacy: Defining Olympic legacies

The analysis focuses on managing Olympic heritage, which is defined by the IOC through the term Olympic *legacy*.

In 2010, the IOC defined the difference between impact and Olympic *legacy*.

The impact is identified as the result of the economic system in the city, while *legacy is recognised* as a positive and long-term element.

These two concepts are related and intersect in different areas:

17   The London 2012 post-Olympic plan is considered to be the best plan in terms of legacy management and organisation.
18   Post-Olympic investments should guarantee a return to the community.
19   For progress on the Olympic legacy, see annexes.

- Cultural, social and political,
- Environmental,
- Economic,
- Urban,
- Sports.

In this sense, the impact is a concrete element that can be direct, indirect, temporary or permanent, and short or long-term. On the other hand, Olympic heritage is used to describe the positive elements and long-term post-Olympic effects on the candidate cities. For this reason, the term *legacy* defines the tangible and intangible benefits of the Olympic Games. The following emphasis on benefits over time has created a lot of confusion among Olympic organisers and researchers.

In addition, the heritage should be considered a negative legacy in case the facilities are abandoned or overestimated. So, as stated by the IOC, it is essential to have a holistic, long-term vision that each year can compare and monitor the positive and negative legacy of the Olympic Project.

3.3.3. Tangible and intangible: Two different ways of looking at the Olympic legacy

Before analysing the different legacies, it is essential to distinguish between two forms of Olympic heritage: tangible *legacy* and intangible *legacy* (Preuss, 2000, 2007, 2021).

As noted, the Olympic legacy can be divided into tangible and intangible *legacies*.

These are some examples of the tangible legacy related to mega-events:

- Sports and non-sports facilities (new construction, reuse, conversion),
- Infrastructures (transport, mobility, etc.),
- Urban fabric and beautification (improvement of the urban areas of the city and works of revaluation of spaces),

- Urban regeneration and rehabilitation,
- Telecommunications,
- Services.

Meanwhile, about intangible legacy, the elements are not so easy for cities to identify and standardise, but can sometimes have a direct and indirect effect that is even more important than tangible factors:

- National or regional pride,
- Improved policies and practices,
- New and enhanced skills and knowledge of the works,
- Changes in attitude,
- Local governance,
- New application methodologies in construction, event management and negotiations,
- Worldwide recognition of the city/country,
- Olympic education,
- Rediscovery of national culture,
- Environmental awareness and sensitisation.

For this reason, a short-term, temporary or transitory legacy may manifest itself before, during or after the Games. The legacy, therefore, can quickly disappear after the event unless efforts are made to keep it alive through, for example, cultural programming, new environmental legislation, public awareness programming or new applications.

3.3.4. *Cultural, social and political legacy*

Expectations related to the event's organisation generate different outcomes in the community that are complex to measure. The promotion of Olympic values, the involvement of volunteers, social integration and the construction of new public policies are the most important topics of discussion to define the success or failure of the event in the local community (Lenskyj, 2006). On the other hand, territorial governance and organisation are critical issues for evaluating the future effects of the Olympic event (Raco, 2013). Significant events generated new conflicts in the community that did not manifest themselves during the development phase of the Olympic Project (Segre & Scamuzzi, 2004).

Cultural, social and political legacies can be identified through the following indicators:

- Individual freedoms,
- Dialogue between different religions and ethnic groups,
- Integration of people with disabilities,
- Fight against exclusion,
- Security and political dialogue,
- Education,
- Rights.

The Olympic Games can also change behaviour and attitudes brought about by the inspiration of the Olympic event.

- Volunteering,
- Cultural pride,
- Increased self-esteem,
- Awareness raising and empowerment,
- Community empowerment and civic awareness,
- Inclusion,
- Interest in and knowledge of the country's history and culture,
- Interest in and understanding of art and gastronomy.

In this sense, new forms of governance and practical tools during the different phases of bidding and post-Olympic planning can become catalysts for implementing further actions of interest to the local community and host city. Therefore, a change in the attitude of local authorities can enable the development of a new form of governance, participation and territorial cooperation.

Moreover, the use of the Olympic brand temporarily projects an image of excellence and inspiration to the world. However, promoting the city is a crucial element, especially in the run-up to the event, reinforcing the association between the city and the collective imagination. Thanks to this new temporary image, host cities can acquire new values and new forms of inspiration for implementing new social projects and other local initiatives. For this reason, the local community can be seen as the subject of heritage protection.

3.3.5. The environmental legacy

The environmental legacies of the Olympics can be broadly seen in the following categories:

- Improvement, implementation and preservation of the environment,
- Design and construction of environmentally friendly Olympic facilities,
- Promotion of new environmental management practices and standards,
- Demonstration of new environmentally friendly techniques and technologies,
- New approaches to the procurement of sustainable processes and contractual sponsorship requirements.

The first two categories constitute tangible legacies that primarily benefit local communities. The others may also include tangible legacies adopted and applied further in the field, although their specific impact is often more difficult to calculate. It should be noted that consideration of the environment only entered the agenda of the Olympic movement in 1999 during the IOC's World Conference on Sport and the Environment (Furrer, 2002). The Olympic movement in Seoul adopted Agenda 21, which included the points identified at the 1992 Rio conference. Therefore, Agenda 21 set out a series of actions to promote the concept of sustainable development and respect for the environment in the organisation of the Olympic event. This document needs to be revised and proposes challenges for preparing and organising the Olympics; it will be up to the Olympic cities to interpret the guidelines autonomously. For example, Lillehammer (1994) is recognised as the most sustainable Winter Olympics in history today[20]. Subsequently, Sydney was the first Olympic city to include specific environmental guidelines in its initial bid. Another example of good practice is the approach the Turin Organising Committee (2006) adopted to evaluate the Olympic plan. The use of the Environmental Impact Assessment (EIA), through continuous monitoring in the different phases before, during and after the Olympic event, has allowed the organisers to be promoters of a practice that is nowadays mandatory to host the Olympic event.

From the list of sustainability indicators and processes observed in Olympic history, we have been able to highlight the following activities:

- Sydney 2000→ Specific audits for environmental management,
- Turin 2006→ ISO14000, EMAS and EIA,
- Vancouver 2010→ SMRS (Sustainability Management and Reporting System),
- London 2012→ Specific audits for environmental management.

3.3.6. *The economic legacy*

The economic legacy is the most attractive for the organisation, as is the evaluation of the impact of the Olympic event by public administrations. The event's financial success is the legacy that is of most interest to the Olympic movement for the redistribution of benefits and the promotion of an economically sustainable event. In addition, the economic success of the event implies an increase in interest from sponsors and the media. However, proper economic planning can benefit the host city in reviving the service economy and repositioning itself in the framework of international financial flows. For this reason, the economic and other legacies

---

20   Planning temporary facilities and structures through wood has reduced land consumption and sustainable promotion of the Olympic event worldwide.

must be planned, supported and implemented in the post-Olympic period. In such a way, their effects continue over time.

Throughout Olympic history, we have seen how the Olympic brand, starting with Los Angeles in 1984, has been incorporated into a territorial marketing strategy to attract tourists, companies and new events to the area. The Los Angeles (1984) and Atlanta (1996) Olympics are defined as the best events in terms of economic performance and performance obtained through repositioning the contemporary metropolis in a framework of global cities. However, the Barcelona edition (1992), developed through a mixed economy, was a great economic success that transformed the city into a contemporary metropolis and a popular destination for tourists and international companies[21].

On the other hand, Sydney (2000) is considered a success story of territorial marketing, having a tangible impact on the perception of the city as a tourist venue for cultural and sporting events. The Sydney Organising Committee aimed to re-launch Sydney's tourism, as there were severe accessibility constraints to the mainland (Cashman & Hughes, 1998). Sydney's bid was developed through the collaboration of the National Tourism Agency, which, for the first time, developed a territorial marketing strategy involving foreign journalists, companies and international tour operators. The synergy of different *stakeholders* allowed the whole world to get to know a continent that, until then, had not been a significant tourist destination.

Preuss (2000), looking at the macroeconomic level of the Olympics, argues that the event can lead to an increase in the demand for consumer goods and the establishment of new businesses. He also analyses how the demand for employment is a temporary element of the bidding phases due to the construction of the Olympic works.

Therefore, the same author proposes some indicators for the evaluation of the economic legacy of the Olympic Project:

- Consumption,
- Unemployment/employment rates,

21   Today, the city of Barcelona continues to invest and implement financial resources to explore the service economy to be inclusive, participatory and communicative.

- Permanent employment,
- The arrival of new companies,
- Several tourists,
- Several nights in hotels,
- Several events attracted me.

Preuss (2000), in his study, states that Olympic cities can suffer from a temporary displacement effect. The effect occurs when, in a short time, economic resources shift from one sector to another. This tends to happen in industrial cities that have turned resources towards tourism services. For this reason, the increase in demand requires an increase in the productivity of goods and services and long-term investments. During the Olympic event, rental prices tend to rise, which can lead to real estate speculation over time[22].

On the other hand, the temporary price increase can lead to a rise in the cost of Olympic construction work, leading to an overestimation and future abandonment of Olympic facilities[23]. Therefore, in 2018, the IOC introduced new selection criteria for candidate cities to reduce the size of Olympic projects in future cities.

3.3.7. The urban legacy

Meanwhile, the analysis of the urban *legacy* of the Olympics can be identified in three main categories:

- Urban renewal,
- Construction of new urban areas,
- Improvement of infrastructural networks.

22 For example, in Barcelona, after the Olympic Games, the value of real estate in 1993 was 300 per cent higher than in 1992.
23 The risk of failure is relatively high if we look at the experiences of Sochi (2014), Rio de Janeiro (2016) and PyeongChang (2018).

Urban renewal represents a unique opportunity to renovate candidate cities and beautify urban areas that will be part of the Olympic Project. The IOC's objective is to encourage the creation of a more attractive place to live in the city with a higher quality of life through better spatial conditions and to set long-term goals. In this sense, the renovation and beautification of spaces are fundamental factors in constructing a tourist city accessible to all (at a global level).

On the other hand, the construction of new urban areas, the evolution of transport lines and the expansion of the service sector are the main elements that explain the appearance of large areas of abandoned, unproductive industrial land in strategic areas for the new future of cities. The connection between the new areas will be fundamental for implementing the infrastructure project and improving viability in the host city. The need for infrastructure to host the Olympic Games meets the opportunity to mobilise the necessary forces for the development of new neighbourhoods and new suburbs.

Moreover, as explained in the specific section, improving infrastructure networks can determine the success of Olympic Games organisations and requires reliable, fast and safe transport[24].

The process of urban transformation and, consequently, the urban legacy of the Olympic Games can be seen as a catalyst for urban change in the post-Olympic phase, capable of stimulating the creativity of territorial planning and becoming a sort for future projects. However, urban transformations mobilise public and private funding in the post-event period. The city can improve its decision-making positions and determine new land changes for the post-Olympic requirements. For this reason, consideration of the urban legacy should be added to the ongoing planning of host cities, reducing the risk of ephemeral projects and the production of obsolete structures in a post-Olympic phase (della Sala, 2022).

However, having observed the experiences of Turin, Vancouver, and London, the post-Olympic phase requires organisational planning to

24   For example, in 2008, Beijing invested twenty billion in the transport and road infrastructure sectors alone. Over time, we have observed different infrastructure projects that underline the importance of the urban legacy.

achieve the objectives set in the bidding phase. On the other hand, the urban legacy is one of the most complex elements to realise and shape the overall Project of the host cities. Regarding culture and political time, over time, some Olympic projects have resulted in a negative urban legacy that remains a problem for Olympic cities today[25]. Thus, urban legacy is strongly linked to each specific urban context, and for this reason, the Olympic Games have, in some editions, underestimated the urban results achieved[26].

3.3.8. *The sporting legacy*

The sporting legacy in the candidate cities can be observed through sports infrastructures, new permanent venues, the improvement of existing facilities, the construction and rehabilitation of new sports venues and the promotion of sports equipment. On the other hand, the sports legacy for the population can be observed through citizen participation in sports events and sports practice. Regarding sporting legacy, organisers should ensure that a facility's use and configuration take into account the experience and form of each city in consideration of its incorporation into a circuit of sporting facilities. In addition, consultation with residents and potential future users will provide insight into the demands and needs of each user in the post-Olympic period. Finally, reducing the size[27]. Building a facility after the Olympic period by creating a mix of structures is only an introduction to the temporary facilities we will see in the future (della Sala, 2022).

---

25 Turin's Olympic Village in the Lingotto area is one of the works that, to this day, has not left any positive urban legacy for the neighbourhood and the entire southern area of the city.

26 The editions of Rome, Tokyo, Seoul, Barcelona, and Turin were supported by ten-year development plans where the event was only a pretext for implementing the urban transformations envisaged above.

27 The new Olympic projects imply a post-Olympic redefinition in search of a decrease in the number of seats. However, post-Olympic redefinition will entail new funding.

3.3.9. Consideration of the Olympic legacy

During the evaluation phase of the Olympic cities, the IOC needs to fully consider the city typology, citizen participation, territorial acceptance, and the implementation of new strategies to change the city image[28]. However, the IOC should implement the bidding process, supporting candidate cities throughout their cycle. Thus, the Olympic legacy has a dual purpose: one for the city and its citizens and one for the Olympic movement. This division of objectives requires guidelines for planning and implementing detailed programmes that can assist the citizens and the city in choosing the Olympic Project. Local authorities and the population cannot be excluded from the bidding process and the planning of the Olympic Games. Undoubtedly, the success of the Olympic Project provides an intangible image return for the entire Olympic movement (della Sala, 2022). Inevitably, the city's image will be projected on the international event circuits, introducing a new model to be pursued. Over time, many Olympic cities have served as a model for future candidate cities[29]. Therefore, these cities' success was manifested in their Olympic plan, the transformation of the cities' image, and the economic and financial development of the territory in the post-Olympic phase. For this reason, the editions that have been successful and will be successful in the future are destined to serve as an Olympic model, influencing the future projects of the new Olympic cities. On the other hand, equally important are the negative editions in terms of urban transformation[30]. On the occasion of negative editions, the IOC has had to modify the process of city allocation so that the Olympic Games can always represent a clean and dynamic product for sponsors and the media. For this reason, promoting an

28 Participatory planning between citizens and local authorities should be integrated into the Olympic Project at all stages.
29 Rome, Mexico, Munich, Los Angeles, Barcelona, Sydney, Turin, London, and Tokyo are just some projects developed through a reference model or by creating a new spatial development model for the city.
30 Editions such as Montreal, Athens, Sochi, and Rio have promoted and publicised instability, causing a lack of credibility for the Olympic movement.

Olympic city and an urban model to be pursued can catalyse supra-local transformations that support the intangible knowledge of the citizenry. The social transformations through the promotion of the city and its image can remain a "Know-How" of the Organising Committee, which in the post-Olympic phase will support the future strategies of the city regarding the organisation and planning modalities of the mega-events.

The Olympic *legacy* is very imprecise about its dynamism over time. The Olympic *legacy*, tangible or intangible, direct or indirect, modifies its space and temporal performance. Therefore, the change of scale of the post-Olympic Project allows for introducing impact or outcome terms. Some researchers, such as Cashman (2002), prefer to define *legacy* as *hardware* (facilities and infrastructure) and *software* (culture, image and identity). The above division of tangible and intangible resources allows us to reflect on the importance of defining the Olympic Project so that socio-economic and morphological problems do not arise in the Olympic territory. In line with the contributions of Rafestin (1981) and Turco (1988), territorialisation can be defined as the production of territory, a territory considered as a space produced by the action of all the actors who carry out projects on the territory. This territorial space can be defined as a space where energy and human capital are applied (Raffestin, 1981).

Moreover, territorialisation in the Olympic Games starts from the bidding phase, transforming itself during the organisational phases of the event until it becomes a process of de-territorialisation in the post-Olympic period. In this last phase, many works have been dismantled or abandoned. Meanwhile, the next phase of territorialisation is defined through the city's legacy plan, which transforms the heritage into a tangible heritage for the city and its citizens. According to Turco, territorialisation, observed as the production of new spatial territory, fits into the ordinary transformation dynamics of cities through a T-R-D cycle that can be analysed as different acts of territorialisation that can be defined through the definition of three categories: Naming, Reification and Structuring (Turco, 1988). The naming of the Olympic territory is related to the control of the symbols of the territory: Olympic stadium, Olympic Plaza, Olympic Village, Olympic track, etc. Over the years, these symbols will inevitably change in form and specific weight of their involvement and application time. Throughout Olympic history, they were sometimes renamed in the aftermath of the event to exploit the brand of the sponsor offering the most long-term funding. In this sense, Olympic facilities and works in the post-Olympic period will inevitably change their value and significance in the city. Therefore, constructing defined Olympic strategies ensures the development of a city brand that will inevitably be promoted to the world. For this reason, the transformation of temporary sites during the Olympic event can be identified as the first phase of territorial production.

After London 2012, the *legacy issue* became a key element and should be seen as the last phase of the Olympic cycle, but the de-territorialisation phase should not include a new phase of re-territorialisation.

The de-territorialisation phase can lead to the following transformations in the territory:

- Dismantling.
- Reuse.
- Abandonment.

Therefore, the most significant risks of the IOC and the candidate cities may lead to excessive Olympic territorialisation, as in the post-Olympic period, they may be transformed into abandoned structures (della Sala, 2022).

It is argued that over-production and a dispersive project can lead to a territorial deficit, housing occupation, reuse of spaces and transformations in mountain areas.

The Olympic Villages

ABSTRACT

After observing Olympic town planning, the following contribution will analyse Olympic accommodation in view of the different obligations and requirements that candidate cities must fulfil according to IOC reports. Subsequently, through the conclusions of the 1996 Olympic Symposium, the chapter will introduce some considerations on the evolution of the Summer and Winter Olympic Villages. In line with the contribution of Muñoz (1996), the different solutions adopted by the host cities will be analysed. Therefore, the Olympic Villages, in particular, will be explored through the processes of construction and reuse of accommodation over time with the contribution of della Sala (2022).

4.1. Planning the Olympic Village

4.1.1. Introduction to the Olympic Villages

> The Olympic Games through the candidate cities represent the image of the strategy for the promotion of the space with the achievement of a competitive advantage over other cities. (Whitson & Macinthos, 1996)

Since the second half of the nineteenth century, candidate cities have discovered mega-events and their potential. Therefore, cities began to use the event to promote their image to the world, accelerating the process of globalisation. According to Hiller (2000, 2003), an event organised on a large scale can be considered a mega-event. However, only if the event has a significant and permanent urban effect on the urban fabric can it be defined as a mega-event. Moreover, the mega-event can induce different physical transformations that amplify the spatial impact and accelerate ordinary transformation processes Essex and Chalkley (1998).

Including the Olympic event within the municipal urban agenda will inevitably change priorities within the overall transformation plan. Furthermore, future space modification or alteration works will promote the Olympic Games' urban legacy (Hiller, 2014). As noted in Chapter 2, mega-events imply a rethinking of the priorities of regular urban processes, requiring long-term and large-scale preparation in order to optimise economic resources.

Therefore, Olympic urbanism in general and Olympic Villages in particular, considering the processes of construction and reuse of the Olympic Village, represent a specific case of urban transformation (Muñoz, 1996). The Olympic Village is regarded as the centre of the Olympic project both for its functionality during the event and, above all, for its subsequent use (Muñoz, 1996). In this way, the Olympic event makes it possible to observe the urban evolution of the host cities through the creation of new urban areas and the restructuring of spaces within the urban fabric.

Research on Olympic Villages involves the study of the city, planning and punctual transformation processes for the temporary accommodation of athletes. Therefore, with regard to the theoretical concept of the Olympic Village, it is essential to introduce the origin of Baron Pierre de Coubertin's original thinking. The idea of creating a "modern Olympia" was openly discussed between the Baron and a group of architects as early as 1910. At that time, the Olympic Village was defined as a unitary complex organised in different locations for the celebration of the Games. The Village was inspired by internationalism and aspirations for world peace, which was characteristic of the thinking of the European intelligentsia in the first half of the twentieth century (Greslery, 1994). Therefore, the Baron's idea was to create a space that had its own territorial dimension by introducing elements such as sport and education to achieve multidimensional goals. From this point of view, Coubertin's proposal had many aspects in common with contemporary ideas, such as the "international city" conceived by architect Ernest Hébrard as early as 1910[1]. If the international city was defined as

---

1   Muñoz, F. (1996). *Historical evolution and urban planning typology of Olympic Villages*, , p. 28.

the new capital of peace and thought, Coubertin's Olympic city could be defined as the capital of peace and sport (Muñoz, 1996).

Therefore, in view of the number of athletes, officials and visitors, the IOC was forced to consider the issue of accommodation as an organisational and decision-making priority in terms of electing host cities. Undoubtedly, the availability of hotels by host cities was a primary concern of the IOC during the first two decades of the twentieth century.

However, at that time, the IOC was undefined (Muñoz, 1996), especially in terms of the budgets of the different countries, so it was not easy to manage accommodation independently. The first solution adopted by the IOC was to allocate the event only to cities with hotel availability and negotiating capacity. However, the accommodation situation was characterised by total improvisation, so much so that some countries used boats as a means of transport and accommodation for their delegations[2]. As the number of participants and spectators increased, the Organising Committee had to work to find other places that could be temporarily converted into accommodation (such as hospitals, schools, military camps and boats). The first decade of the twentieth century was characterised by the emergency of finding accommodation for Olympic athletes. However, the regulation of the event and the temporary provision of the first Olympic Village in Paris in 1924 provided a clear signal to introduce a new debate on Olympic housing. Therefore, the first phenomenon observed in the first Olympic Village in Paris in 1924 is undoubtedly an emergency (Muñoz, 2006). After Paris 1924, the temporary need to host Olympic athletes meant that the IOC had to make decisions to guarantee accommodation for the duration of the event and for all participants. It is crucial to emphasise that the first Olympic Village in Paris was proposed on a military urban scheme and represented by artefacts such as barracks and huts. In addition, its construction timing allowed it to occupy an area close to the Olympic Stadium to provide essential services and favour Olympic athletes. However, the Paris Village had little in common with the first Olympic Village built for the 1932 Los Angeles event. The Los Angeles Village of

2   The issue of travel and accommodation costs for delegations will be one of the main topics of discussion in connection with the increase in the number of participants.

1932 was to be the promoter of a new suburban model inspired by the new temporal urbanisation and modular construction techniques. As in Paris, the Los Angeles Olympic Village will be built temporarily in a suburban area with ample training and recreational space for the Olympic athletes.

However, during the Berlin Congress 1930, IOC members initiated a debate to promote a new permanent housing solution. Zack Farmer, the member in charge of operations, promised a new urban model to advertise accommodation, providing a solution that included two dollars a day for food[3]. Thus, in 1936, Berlin would be the first city in the world to provide a permanent urban planning model that included housing, services, transport, sports stadiums, and open spaces. From this historic moment, the construction of the Olympic Village would begin to take the form of a unique construction site with a tremendous physical impact on the host city. As stated in the official OCOG Report of 1936, the Berlin Organising Committee's desire was to replicate the Olympic Village in Los Angeles to emphasise and replicate the modern city of Elis (OCOG, 1937).

Therefore, the Organising Committee proposed a permanent solution in a suburban area, as previously observed in Los Angeles. The area identified was the Döberitz military camp, some 21 km from the Olympic site. The women, however, as in Los Angeles in 1932, were accommodated in different locations. The Berlin urban complex and the single-family villas in Los Angeles were the typologies that would promote and inspire a housing model laying the foundations for buildings throughout the twentieth century. Housing defined and shaped a new image of the "Olympic city" from this historic moment. The construction of housing with the inclusion of sports facilities introduced a new imagery of architectural ensemble for implementing new urban strategies. After the 1936 edition in Berlin, the conception of the Olympic residence as something more than a temporary place of accommodation was born. A residence encompassing different aspects, meanings, spaces and services that began to evolve through the definition of other proposals for the realisation and redefinition of spaces

---

3   The American offer of accommodation, meals and use of local transport was hard to refuse, and therefore, the Village of Los Angeles will become an inspirational model for future candidate cities.

beyond those specific to sport. Therefore, during the twentieth century and more markedly in the twenty-first century, the Olympic Village has become the fundamental element in promoting a structural modernisation of cities, which in some cases has become a model for developing housing within host cities. The observation of successful models such as Rome (1960), Munich (1972), Barcelona (1992), Sydney (2000), Vancouver (2010) and London (2012) implies a rethinking of the housing strategies of future cities. Undoubtedly, the Olympic Village, like the stadium and the Olympic pool for the summer edition, is the structure that constitutes the structural heritage of the Olympic Games in the host city. An urban heritage that, considering each host country's cultural, social, political, economic and sporting history, can become a crucial element in creating new socio-economic transformations. Undoubtedly, the Olympic event and the Olympic Village, starting in Rome in 1960, will play a vital role in restructuring urban space and the dimension of the Olympic event[4]. The 1960 Rome edition will introduce new elements that will imply a rethinking of the design scale. In the editions of Oslo, Rome, Mexico, Grenoble, Munich, Barcelona, Sydney, Turin, Vancouver and London, we have observed how the Olympic Villages can become an active and dynamic asset that continues to catalyse new transformations.

Therefore, the Olympic Village's potential should be evaluated by observing other elements that can benefit the city and the future of our territories. The Olympic Village is the cornerstone of the renewal project through the Olympic event. The urban style, formal language, choice of materials and application of new building technologies, supported by infrastructural transformations, represent an intangible possibility for candidate cities. However, the possibilities and choices must consider the city's existing plans and future projects. As we will observe within Chapter 5, in the history of the Olympic Games, the types of Olympic Village adopted by candidate cities can be analysed through different permanent and temporary models.

---

4   In 1960, Rome was the first Olympic city to use the Olympic event as a catalyst for other urban and infrastructural transformations proposed by the city's post-war reconstruction.

The construction of new housing and land consumption can not be considered only to fulfil IOC obligations. Cities should consider the Olympic Village as an element that can be integrated within a new urban development philosophy that, through sport, can promote a healthy lifestyle within our cities.

Therefore, only by considering the real needs of citizens will the Olympic Villa project be able to meet the expectations of the host community. Otherwise, temporary solutions are the best measures to avoid jeopardising the future of the candidate cities. An abandoned housing complex of this size, as observed for the Olympic Villages in Athens, Turin, Sochi, Rio and PyeongChang, still compromises the future of the host cities. The neglect is a consequence of choosing a permanent model not included in the post-Olympic housing planning. Therefore, these examples of structures built and then abandoned allow us to reflect on the importance of urban planning in not permanently compromising the territory of host cities.

Over time, we have observed different reuse or renovation projects of urban spaces that integrated Olympic housing without compromising urban form and scale. However, awareness of the impact of the Olympic Village in central areas has introduced new models of mixed housing and services. The construction of new housing in the main fabric of the city will inevitably lead to changes in the services available with a consequent increase in prices, accelerating the processes of gentrification in the new Olympic quarter.

Moreover, in some editions such as Sydney, Athens, Beijing and Rio, the identification of land has led to a displacement of people for the construction of the Olympic Village. At the same time, the transformation of the areas in Beijing caused a change in the value of the land with the subsequent change of use in the post-Olympic phase (Zou, 2015).

Land-use change should be contemplated beforehand and included in a master plan with objectives to transform cities' socio-urban and economic fabric. Undoubtedly, changes in the euro value of land can introduce phenomena related to real estate speculation and gentrification. However, on the other hand, the Olympic Games adopting polycentric spatial patterns will contribute to raising changes in different areas of the territory. Therefore, the spatial dimension of the event is a crucial factor that cities must carefully observe in order not to make choices that could compromise the future development of our territories.

However, according to Millet (1997), it is complex to develop the Olympic Village through mixed financing in some cities, so cities will have to justify the construction of accommodation in other ways. Therefore, the choice of different locations can improve the distribution of facilities in the territory and reduce the impact on the host city. Millet (1997) and Muñoz (1996) argue that it is not possible to identify a joint development model in the history of the Olympic Village.

Therefore, the model and size of the Olympic Village will depend solely on its urban context and current plans.

Millet (1997), in his analysis, identifies four generic cases that represent the planning of the Olympic Village:

- *Zero impact* Where the Olympic effect does not cause much structural change. (Los Angeles, Calgary, Atlanta and Salt Lake are examples of the use of student accommodation).
- *Urban spread* The Munich Olympic Village of 1972 respected the general strategy of the city's master plan, developing new neighbourhoods with access to infrastructure and parks and using the Olympic Village as a model.
- *Urban renewal*→ Rome, Tokyo, Barcelona and Turin are the most important examples. The Olympic Village was developed with more qualitative than quantitative aspects in mind, allowing a temporal revaluation of the area.
- *Mixed option*→ In Seoul 88, we can see how the Olympic Village was in the middle of relocating factories on its territory (Millet, 1996).

After introducing Olympic Villages, we will examine the construction standards proposed by the IOC for the provision of accommodation in the following sections. Subsequently, we will observe Olympic accommodation's different typologies, spatial models and formal languages in the summer and winter editions. By following the different evolutionary phases of Olympic housing, the reader can provide some insights into the future new projects to be planned by the candidate cities.

4.1.2. The Olympic Village symposium

The 1996 Lausanne Symposium on Olympic Villages is regarded as the first international scholarly discussion. The symposium was organised on the initiative of the Centre d'Estudis Olympics, the International Olympic Committee and the Olympic Museum in Lausanne.

Today, the 1996 symposium is regarded as the only time for reflection on the projects implemented and the development of the subject at an academic level.

Thanks to the contribution of international experts, the symposium showed that the Olympic Villages raised many important questions about the relationship between the Olympic Villages and the Olympic Games:

- Planning/market.
- Redevelopment/extension.
- Public sector/private initiative.
- Social policies/personal profit.
- Integration/segregation.
- Innovation/tradition.
- Environmental concerns/economic growth.
- Utopia/reality.

Looking at the following elements and relationships, we can state that the planning and choice of the Olympic Village area can be decisive for the host city. Architectural design, redevelopment of the area and intensification of services will be some of the most influential elements in determining the functioning of the Village in the post-Olympic period. In addition, the symposium will emphasise the importance of Olympic accommodation as a central place for the transfer of the Olympic experience among all cultures of the world. During the symposium, the importance of designing and considering the Olympic Village as a central place where all the world's athletes will live together for the duration of the event was emphasised. In this way, the accommodation will contribute to the promotion of the noble ideals of understanding and friendship among participants from all countries of the world.

Today, the Olympic Villages are expected to accommodate the following number of participants depending on the edition:

- Summer: approximately 20,000 people
- Winter: about 5,000 people

In addition, the following types of housing must be planned to meet IOC requirements.

- Development of new housing in the host city
- Use existing housing, such as hotels, student dormitories, resorts, barracks, etc.
- Temporary structures such as bungalows or demountable caravans.

Therefore, the symposium seeks to offer a review of the different phenomena observed up to 1996, proposing a reflection for the future debate on Olympic accommodation (Moragas, 1996). During the symposium, Muñoz (1996), for the first time, introduces a new observation of the different solutions adopted by the host cities, analysing the other urban models and formal language that have permanently marked the history of the summer edition. Throughout the Olympic Villages' history, we have observed different solutions in the typology and integration of permanent structures into the urban fabric of the host cities. As we will see in the next section, the other solutions adopted by the host cities will allow us to reflect on the evolutionary stages of Olympic construction.

During the symposium, reflecting on the importance of the transport system during the Olympic event, the IOC will introduce for the first time the following exclusive feasibility typologies for the athletes' routes during the event:

- Within the Village (zero emission)
- From the Village to the competition sites
- From the Village to the city centre
- Between the airport and the Village

Creating the new exclusive routes will entail a new rethink in the design and planning of the Olympic Village. After Atlanta 1996, the IOC wanted to intervene in order not to jeopardise the execution of the Olympic event due to delays in sports competitions. Therefore, regarding the Olympic Village, the IOC obliges the organising Committee to set up a steering committee four years before the Olympic event to plan and manage the operations of the Olympic Village (IOC, 1996). Furthermore, the management of the Village must be implemented from its first conception. This includes the operational arrangements and strategic planning to ensure its entire operation during the hosting of the event.

The IOC 1996 identifies the following steps for planning the design of Olympic accommodation:

- Planning Assumptions
- Identify and acquire
- Block planning
- Permanent design and construction
- Temporary design, construction and outfitting

In addition, factors to be considered in determining the permanent design should be:

- Physical constraints of the Olympic Village site
- Requirements defined in the Olympic Village Technical Manual
- Requirements specified during the planning of the block
- Use of the permanent buildings after the Games
- Infrastructure requirements (sewage, waste, water, energy).

Subsequently, the IOC, recognising that funding for the permanent construction of the Olympic Village generally comes from government sources, identifies some possibilities for future host cities.

- Private investors
- Public companies
- Future owners of the site
- The construction company itself

In conclusion, during reflection, it was noted that the Olympic Village must also mandatorily include other facilities and services for international athletes. Furthermore, in the next paragraph, we will see the standards and obligations to be met for the constriction. Like the other Olympic facilities, the Olympic Village must be built according to the IOC guides and technical manuals, following the guidelines for the construction and operation of the entire Olympic Village area. In addition, the Olympic Village must include the sports facilities. Therefore, the host cities must follow the instructions in the specific technical manuals. About the sports facilities to be included in the Olympic Village, the IOC makes the following differentiation:

- Winter: Gymnasium, men's and women's changing rooms
- Summer: Gymnasium, jogging track within the Olympic Village (1 km minimum), saunas, and changing rooms.

In observance of the topics covered, the 1996 Olympic Symposium is a starting point for a multidisciplinary reflection on the study of summer and winter Olympic Villages. However, as noted in the doctoral research results, the organisation of a new symposium is of fundamental importance for observing the new urban phenomena that have emerged in the twenty-first century.

4.1.3. Main requirements

In the following section, we will look at the obligations required by the IOC regarding constructing the Olympic Village in the host city.

- The predisposition of the areas is as follows: Residential Area (athletes' accommodation) and International Area (services, shops and cultural facilities). The Olympic Village must remain open 24 hours a day for authorised persons.
- Dimension: Summer (20,000 athletes), Winter (5,000 athletes)

- Typology: New development, existing dwellings, temporary dwellings
- Additional staff 1,550 (1,000 rooms) and 800 for Winter Games (300 rooms)
- Dimension of accommodation: for two (2) athletes a minimum size of 15mq2 and a private bathroom must be respected. Meanwhile, for officials, eight-person flats can be provided with at least three (3) toilets and three (3) showers.

The organisation of the area international

- Administrative headquarters of the delegations: Summer Olympic Games: 10,000 sq.; Winter Olympic Games: 2,200 sq.
- Restaurants: Summer Olympic Games 5,000–6,000 seats + kitchens, shops = 12,000 sq.; Winter Olympic Games 1,200–1,400 seats + kitchens, shops = 2,800 sq.
- Delegation offices: Summer Olympics: 10,000 sqm; Winter Olympics: 2,500 sqm
- Changing rooms: Summer Olympic Games 6,000 sq. (for 9,000 people); Winter Olympic Games 1,500 sq. (for 2,100 people)
- Training areas: Summer Olympics 30,000 sq.; Winter Olympics 5,000 sq.
- Leisure centre: Summer Olympic Games 2,000 sqm; Winter Olympic Games 700 sqm
- Shopping centre: Summer YOG 2,000 sqm; Winter YOG 500 sqm
- Logistics centre: Summer Olympics 10,000 sq.; Winter Olympics 2,200 sq.

Services inside the Olympic Village

- NOC Service Centre
- NOC Sports Information Centre
- Shopping centre
- Laundry service, dry cleaning

- Office service centre with language and secretarial services on request, photocopiers, – Service station and/or sports equipment repair shop
- TV room with capacity for live or delayed coverage of Olympic events on request
- Games rooms, discotheques and other leisure facilities
- Relaxation facilities such as sauna, swimming pool, etc.
- Cafés and bars
- Photographic laboratories, flower shops, hairdressing salons, beauty salons
- Post Office
- Bank, travel agency and tourist office
- Various religious centres and meditation halls
- Meeting and conference rooms for team use

Key features of the Paralympic Village

- The geographical location of the largest Village
- Topography
- Relationship to the nearest city
- Nature of buildings
- Rooms
- Toilet/bathrooms
- Food service – 24 hours
- Security
- Internal services
- Support services
- Information system
- Circulation in the village/space
- Climatic factors
- Specialised medicine

Annex : Relational Plan of Surface Areas

(Scale 1:10,000)

(The surface areas indicated give the approximate number of m² of raw floor space developed).

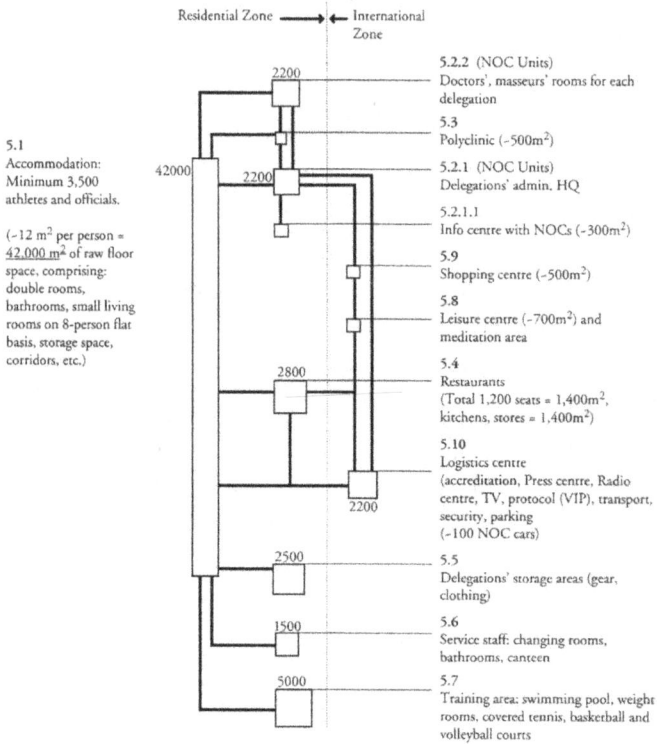

Residential Zone ⟶ ⟵ International Zone

5.1
Accommodation:
Minimum 3,500
athletes and officials.

(~12 m² per person =
42,000 m² of raw floor
space, comprising:
double rooms,
bathrooms, small living
rooms on 8-person flat
basis, storage space,
corridors, etc.)

42000

2200

2200

2800

2200

2500

1500

5000

5.2.2 (NOC Units)
Doctors', masseurs' rooms for each
delegation

5.3
Polyclinic (~500m²)

5.2.1 (NOC Units)
Delegations' admin. HQ

5.2.1.1
Info centre with NOCs (~300m²)

5.9
Shopping centre (~500m²)

5.8
Leisure centre (~700m²) and
meditation area

5.4
Restaurants
(Total 1,200 seats = 1,400m²,
kitchens, stores = 1,400m²)

5.10
Logistics centre
(accreditation, Press centre, Radio
centre, TV, protocol (VIP), transport,
security, parking
(~100 NOC cars)

5.5
Delegations' storage areas (gear,
clothing)

5.6
Service staff: changing rooms,
bathrooms, canteen

5.7
Training area: swimming pool, weight
rooms, covered tennis, basketball and
volleyball courts

Note : The m² indicated correspond to the m² of raw floor space needed (including corridors, etc.)

Figure 6. Dimension of the area of the Winter Olympic Village
(Source: Moragas, 1996)

Annex : Relational Plan of Surface Areas

(Scale 1:10,000)

Games of the Olympiad

(The surface areas indicated give the approximate number of m² of raw floor space developed).

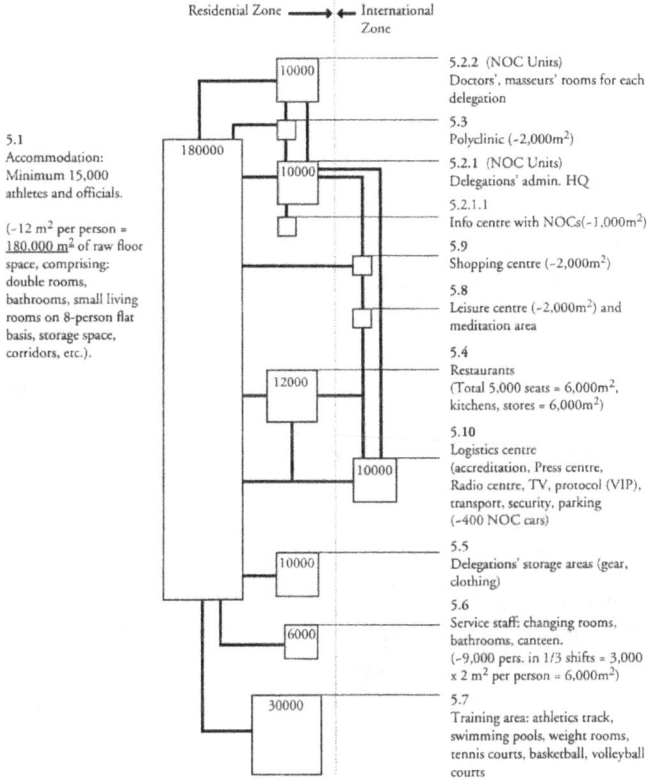

Residential Zone ⟶ ⟵ International Zone

5.1
Accommodation:
Minimum 15,000
athletes and officials.

(~12 m² per person =
180,000 m² of raw floor
space, comprising:
double rooms,
bathrooms, small living
rooms on 8-person flat
basis, storage space,
corridors, etc.).

180000

10000

10000

12000

10000

10000

6000

30000

5.2.2 (NOC Units)
Doctors', masseurs' rooms for each
delegation

5.3
Polyclinic (~2,000m²)

5.2.1 (NOC Units)
Delegations' admin. HQ

5.2.1.1
Info centre with NOCs(~1,000m²)

5.9
Shopping centre (~2,000m²)

5.8
Leisure centre (~2,000m²) and
meditation area

5.4
Restaurants
(Total 5,000 seats = 6,000m²,
kitchens, stores = 6,000m²)

5.10
Logistics centre
(accreditation, Press centre,
Radio centre, TV, protocol (VIP),
transport, security, parking
(~400 NOC cars)

5.5
Delegations' storage areas (gear,
clothing)

5.6
Service staff: changing rooms,
bathrooms, canteen.
(~9,000 pers. in 1/3 shifts = 3,000
x 2 m² per person = 6,000m²)

5.7
Training area: athletics track,
swimming pools, weight rooms,
tennis courts, basketball, volleyball
courts

Note : The m² indicated correspond to the m² of raw floor space needed (including corridors, etc.)

Figure 7. Dimension of the area of the Summer Olympic Village
(Source: Moragas, 1996)

4.1.4. Olympic Village Technical Manual

The Olympic Village Technical Manual[5] It is one of the mandatory documents that host cities must comply with for planning and managing
the Olympic event. As specified in Rule 39 of the Olympic Charter,
Organisers are obliged to create an Olympic Village as a temporary accommodation solution for athletes and Olympic officials. Therefore, the
Olympic Village must be regarded as a strategic place for the transfer of
knowledge that fosters the cultural exchange of participants. However,
certain obligations must be fulfilled to enable the participants' experience.
    The Olympic Village must:

- Operate 24 hours a day
- To be protected from the general public and the media.
- Provide the necessary facilities for athletes and officials.

The objectives of the technical manual on the Olympic Village are:

- Provide applicant and candidate cities with information to develop
  their Olympic Village plans.
- Provide the OCOGs with the structural information to plan and
  build the Olympic Village.
- Provide information on the planning and operational requirements
  of the Paralympic Village.
- Guide an OCOG to design, plan, construct and operate an
  Olympic Village for the Olympic Games, Winter Games and
  Paralympic Games.

However, the requirements of the Olympic Village will inevitably have to
be adapted to the host city and site in question. The requirements of the
Technical Manual are standard requirements that can be modified after a
review with the IOC and interested parties (IOC, 2005b). While the event

---

5   The Olympic Village technical manual is a document that was updated in 2015 and
     is one of the thirty-three technical manuals that host cities are obliged to consider
     for the planning and organisation of the event.

planning is being finalised, the design and construction of the permanent buildings can begin. The design and construction phase of the athletes' accommodation is usually guided by the permanent building authority, which requires frequent reports from the Olympic Village management.

The factors to be considered when determining the permanent design are (IOC, 2015):

- The physical limitations of the Olympic Village;
- The requirements are defined in the IOC Technical Manual on Olympic Village (IOC, 2005d);
- The requirements specified during the planning of the block;
- Use of permanent buildings after the Games;
- Infrastructure requirements (sewerage, waste, water, energy).

Constant communication between the Olympic Village construction authority and the Olympic Village administrative structure is a prerequisite for successful design. The IOC strongly recommends that previous editions' experiences be observed and considered when determining the consumption of water, electricity, gas, logistics and waste to carefully plan the accommodation conversion in the post-Olympic period.

Therefore, the Committee must consider that the Olympic Village requires many more utility resources than an average residential complex. Furthermore, it would be helpful to study the technological requirements of the complex so that they can be incorporated to reduce temporary costs, providing an intangible benefit to the cities in the post-event phase. Finally, consider the need for temporary facilities (e.g. electricity and water) to ensure that the Olympic Village can cope with the additional demand during the event.

4.1.5. Olympic Village obligations

The Olympic Village Technical Manual outlines the following obligations for host cities:

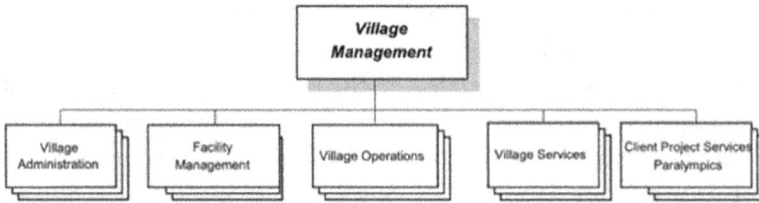

Figure 8. Organisation of the Olympic Village (Source: Own implementation from IOC, 2015)

- In the case of several Olympic Villages, the proposal must be submitted to the IOC for approval;
- The OCOG only has to offer the general layout of the Olympic Village to the IOC for approval;
- Specific visits should be allowed throughout the construction phases of the Olympic Village;
- At the start of the exclusive use, the management of the Olympic Village must carry out a walk-through of the site with the owners;
- Area control points, access control points and vehicle checkpoints must be operational;
- The accommodation must have at most two persons per room.
- No more than four persons per bathroom;
- All NOCs must have adequate accommodation, office space, medical space and storage space;
- The OCOG should provide a media centre in the Village to allow the media a working area in the Olympic Village;
- The OCOG must complete and establish block planning, policies and procedures in the Olympic Village;
- The pre-opening period of the Olympic Games starts seven days before the official opening of the Olympic Village;
- It officially opens fourteen days before the opening ceremony for summer and ten days before for Winter;
- The Olympic Village closes three days after the closing ceremony;
- The OCOG must complete a housing allocation process in the Olympic Village;
- Transport must be operated by the OCOG (IOC, 2005).

4.1.6. Olympic Village managements

Within the Olympic Village Technical Manual, the IOC sets out the following basic organisational structure of the Olympic Village:

Meanwhile, regarding the operational management of the Village, the Committee should direct the following:

- The design of the Olympic Village;
- The strategic planning of the Olympic Village;
- The functioning of the Olympic Village;
- In addition, the Village management must cooperate with;
- THE IOC;
- THE CIP;
- The owners of the Villa;
- Public security;
- Several municipalities/local governments;
- Construction companies/architects;
- Contractors;
- Government agencies, health and safety, and environmental organisations.

In addition, specific visits should be allowed throughout the construction phases of the Olympic Village.

4.1.6.1 Olympic Square

The Olympic Village Plaza[6] (OVP) shall host the following activities and venues:

- Welcome ceremonies for the teams;
- Retail services;
- Recreational services;
- Meeting rooms;
- Village management offices (optional).

---

6   The Olympic Plaza was first introduced in Los Angeles in 1932, and since then, it has become the central venue for the medal celebrations of Olympic athletes.

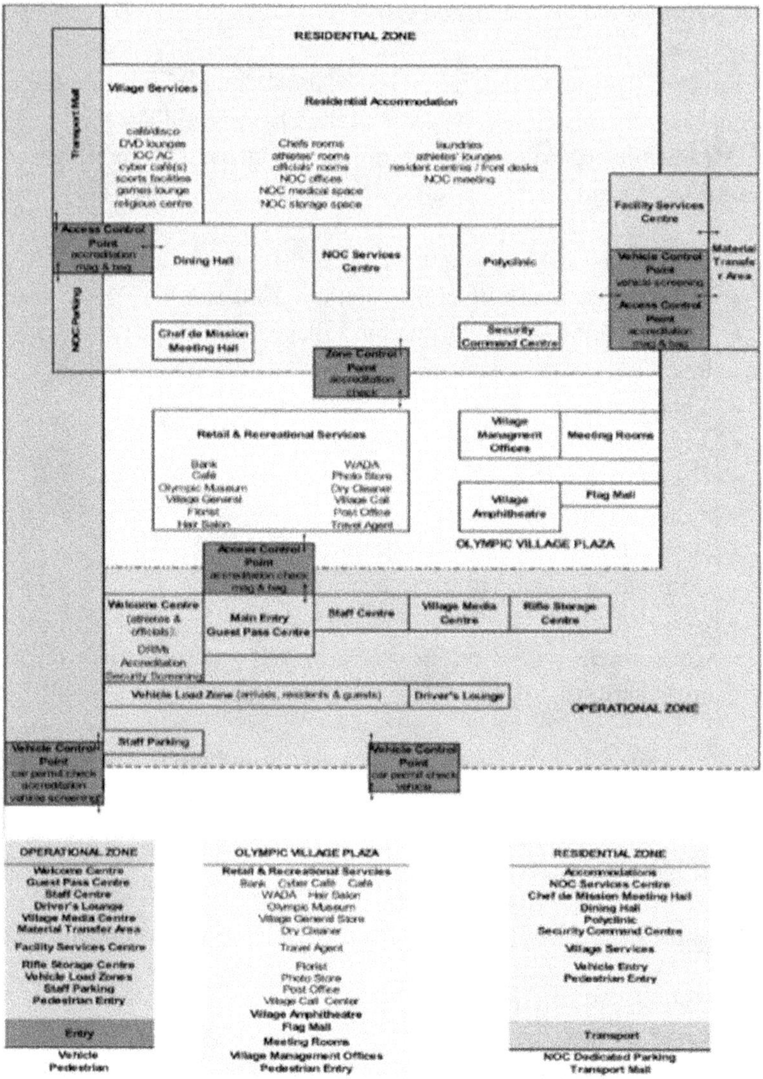

Figure 9. Layout of the Olympic Village area (Source: IOC, 2005)

4.1.7. Olympic Village planning

Operational planning is considered the basis for the functioning of the Olympic Village as it must be organised through four fundamental phases:

1. Strategic planning;
2. Concept planning;
3. Operational planning;
4. Operational readiness.

During this phase, the Olympic Village Department shall complete:

- Strategic scenarios;
- Working meetings;
- Contingency plans and crisis scenarios;
- Actual testing of functions and services;
- Identify the specific roles and responsibilities of all staff, volunteers and contractors.

The Olympic Village is usually the largest construction project of the Olympic Games and involves a considerable investment (IOC, 2005). Different agencies and bodies outside the Organising Committee are involved in the construction phase. Therefore, the ongoing design should be the responsibility of suitable construction companies participating in an open and transparent tender. Furthermore, if the event takes place within the European Union, the awarding process must comply with national and EU legislation. It is crucial for the OCOG that the award decision is based on tangible elements related to post-Olympic planning. External planning implies less control by the IOC over the constructions and the responsible bodies. Therefore, constant communication between the authorities responsible for the permanent construction and the Olympic Village Management is crucial so that the OCOG can constantly monitor the construction and ensure that the works comply with the preliminary planning.

The OCOG shall define the construction requirements for:

- Utilities (e.g. water, electricity, sewerage);
- Technology (e.g. computer network and information system);
- Telecommunications;
- CATV;
- Foundation and soil requirements;
- Heating/cooling requirements; – Safety considerations;
- Main service buildings of the Village; – Access roads and internal roads; – Fences;
- Lifts;
- Lighting;
- Parking;
- Ensuring accuracy and quality of construction;
- Liaison with builders/architects (design team).

The design and temporary construction of the facilities for the construction of the Olympic Village will be determined based on the existing buildings on the site. Therefore, the services must be appropriately located within the permanent structures. The services' location and functions will affect the accommodation's economic efficiency. Only through the design of temporary buildings ancillary to the Olympic Village will it be possible to reduce costs and maximise the benefits of the territorial impact of the event.

In addition, the COJO must define:

- Temporary structures depending on the operational needs of each area;
- Procurement policy and needs;
- Design of temporary structures;
- Terrain, type and size;
- Construction and installation methods;
- Technological and installation requirements;
- Power supply and temperature control requirements;
- Portable toilets;

- Temporary modular constructions;
- Structural tents.

Meanwhile, the winter edition requires many temporary buildings and tensile structures ancillary to the Olympic Village. Buildings and structures will have to be heated constantly during the event.

At the same time, the COJO must:

- Define the perimeter of the site;
- Secure financing before construction;
- Define in more detail the actual perimeter of the Olympic Village site, including security needs and access points.

In the meantime, regarding the perimeter structures of the Olympic Village, the location of the entrances for the pick-up/drop-off of transport vehicles and the parking areas of the Olympic Village, the OCOG should take into account the main service facilities within the Village (e.g. OVP, canteen, shopping centre, etc.). Additionally, the IOC recommends preparing a preliminary study to assess the environmental impact on the site area.

The site assessment should take into account the following elements:

- Toxic waste;
- Contamination at the site;
- Hazardous materials (e.g. asbestos);
- Land quality;
- Soil composition;
- Vegetation and woodland;
- Water drainage and waterways;
- Soil erosion;
- Wildlife habitat;
- The potential impact of construction on the site;
- Emissions/pollution (roads, factories);
- Noise pollution.

Possible environmentally friendly projects for the Olympic Village may include:

- Recycling of materials (e.g. food, waste);
- Reusable energy sources;
- Solar energy (e.g. for heating),
- Minimise wrapping/packaging of deliveries/construction material;
- Recycling of water from the Olympic Village;
- Environmentally friendly Olympic Village vehicles (e.g. natural gas, electric);
- Protection or integration of fauna and vegetation on the site.

Athletes and officials are provided with a unique shuttle service to and from the Village:

- All competition venues;
- All training venues;
- Airport;
- City centre;
- Additional accommodation for officers.

As noted above, the Technical Manual provides for different obligations that may affect the development of an area or an entire host district. However, the IOC does not foresee sanctions or fines for non-implementation and non-compliance. The lack of specific sanctions over the years has led to many housing abandonments that continue to impact the future choices of host cities.

4.1.8. Evaluation criteria for the Olympic Villages

Observing the evaluation criteria of the Olympic Commissions of the IOC working group, the accommodation structures are evaluated through the following three (3) criteria and weightings:

(a) Location 40 per cent.
- Travel distance to competition venues, excluding football and sailing preliminary venues, when outside of the host city.

(b) Concept 40 per cent.
- Number of villages;
- Type of accommodation;
- Available land area;
- Surrounding environment;
- Temporary or permanent;
- Additional accommodation for athletes.

The Village concept will be assigned a feasibility factor based on the likelihood that the proposed projects will be realised.

(c) Bequest 20 per cent of which
- Post-Games use;
- Funding.

As stated earlier, the Olympic Village is one of the most critical sites for athletes during the sports event. The location of the venues has become of paramount importance since Atlanta 1996. After a preliminary phase of the bidding process, the IOC, in the second phase, will assess how the cities will deal with the very complex issues related to the scope and scale of such a project. This project must be supported by operational planning that includes legacy as one of its evaluative elements. The IOC states that most cities have demonstrated a good understanding of the requirements of the Olympic Village, including project legacy (IOC, 2008).

4.2. Urban analysis of the Olympic Villages

A miniature city, replete with modern conveniences and facilities, had magically emerged high in the hills, within sight of the great Olympic Stadium, on top of the modern Mount Olympus, below which lay the modern plains of Elis. (COJO, 1932: 235)

As noted above, Olympic projects can only be analysed by considering our cities' urban history. The theoretical and practical evolution of architecture and the different formal languages used allow us to reflect on the evolution of Olympic Villages in the territories.

The analysis advanced by Muñoz (1996) on the Olympic Villages pays specific attention to four fundamental aspects:

i. Aspects related to the evolution of the architectural idea, the different housing types and the different formal languages used.
ii. Aspects related to the evolution of city plans, from choosing the urban concept model to the basis of the operations adopted.
iii. Aspects related to the conception of the Olympic Village as an urban instrument, from the production of the city's projects to the insertion of the urban context in the post-Olympic period.
iv. Moreover, the change in the economic circuit and the different types of management require a specific section (Muñoz, 1996).

Thus, Muñoz's (1996) study allows us to observe the first classification of the different urban models adopted by the candidate cities for the construction of the Summer Olympic Village, which is: "the garden city, the satellite city, the urban centre and the metropolitan city" (Muñoz, 1996).

The following classification helps us identify some common patterns among the summer editions held so far. However, one of the objectives of the doctoral study (see della Sala, 2022) was to update Muñoz's (1996) contribution for the summer editions and to advance a classification of the urban models adopted for the winter editions.

Therefore, analysing the evolution of the shape and spatial dimension of the Olympic Village, della Sala (2022) advances the following models observed in the winter editions: the satellite city, the mountain centre, the metropolitan city and the cluster.

The following contribution, by analysing the different spatial models proposed by the candidate cities, allows us to reflect in-depth on the evolution of the concept of Olympic accommodation over time and how the Olympic Village has evolved in the two distinct editions. As noted above, the Olympic Village was based on Coubertin's idea of promoting the creation of a sports city capable of fostering cultural exchange among its inhabitants. However, with the construction of the Los Angeles Olympic Village in 1932, the patterns and shape of the accommodation reflected the typical forms and types of each historical moment.

Furthermore, through the classification advanced by Wimmer (1976), three different stages can be observed according to the solutions adopted for the construction of the Olympic Village:

1. The first corresponds to single-family houses or *bungalows* arranged similarly to the housing estates or colonial houses seen in Europe or the United States.
2. The second corresponds to the creation of the community except Melbourne 56 and the integration of wooden modules.
3. The third corresponds to the construction of large complexes on a single module developed in height and identical, in some cases, even with different designs.

Subsequently, Muñoz (1996), analysing the evolution of the form and context of the Olympic Village, will advance the following models and phases that were observed in the summer editions:

1. Olympic Village and urban planning. The utopian content of Olympic urban planning.
2. The garden city and the suburban world. The "inaugural" villages.
3. The satellite city and the city machine. The people of the 1960s.
4. The central city and the accumulation of leisure. The people of the 1970s.
5. The metropolitan city and the central "non-place". The people of the last two decades.

Therefore, as we will observe in the following section, the Olympic Villages, in their history, have had to adapt to various transformations and modifications in order to fit into a long-term urban development framework (della Sala, 2022). However, some cities have had to undertake structural changes to meet new housing needs due to the exponential increase in urban population. The city-village model has been entirely replaced by the region-metropolis model, which continues to expand in size by introducing different types of spatial models. The expansion of the Olympic space has complicated the definition of the services and responsibilities of the public administrations involved in the Olympic project (della Sala, 2022). Consequently, the Olympic Village must be seen as the result of Olympic time in a contextual territory that will have to fit into a permanent physical structure, responding to the specific housing needs of each host site. The following section will analyse the spatial models adopted so far for constructing the Olympic Village in the host territory.

4.2.1. *Spatial models of the Olympic Villages*

Over time, different spatial models were proposed for the location of the Olympic Villages in the host city. Moreover, from 1924, with the establishment of the winter edition, the organisation of the Olympic Villages was transformed, and new elements and functions were acquired.

During the twentieth century, the evolution of the Olympic event introduced new ideas that led to a rethinking of the event model and the temporary accommodation of athletes.

The editions of the Olympic Games from 1896 to 1924 were organised based on a spatial model that placed the Olympic Stadium at the centre of the project. While starting with the 1932 Los Angeles edition, the event allowed the world to observe a new peripheral spatial form, and the Olympic Village became the central element of Olympic urban planning in the host territory. Therefore, the 1932 project must be regarded as the first temporary sports district built and still used today. Los Angeles was a source of inspiration for the construction of future accommodations for Olympic athletes. Thus, in the Berlin Report of 1936, the Organising Committee emphasised the importance of developing an Olympic Village with the same philosophy and form as the model observed in previous Games. The Berlin edition can be regarded as the first edition that drew inspiration from past editions. It proposed developing a permanent accommodation model that could be replicated over time.

Spatial dimension of Olympic Villages

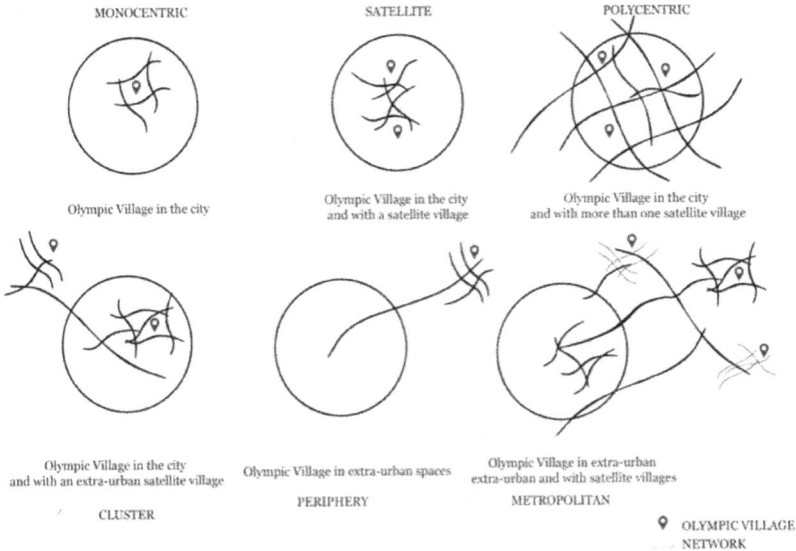

MONOCENTRIC

SATELLITE

POLYCENTRIC

Olympic Village in the city

Olympic Village in the city
and with a satellite village

Olympic Village in the city
and with more than one satellite village

Olympic Village in the city
and with an extra-urban satellite village

Olympic Village in extra-urban spaces

Olympic Village in extra-urban
extra-urban and with satellite villages

CLUSTER

PERIPHERY

METROPOLITAN

♀ OLYMPIC VILLAGE
NETWORK

Figure 10. Spatial dimension of the Olympic Villages (Source: Own implementation)

Starting from this moment, the cities awarded the summer edition will begin to propose and develop urban projects that include the sports facilities and services surrounding them.

Meanwhile, for the winter edition, the first Olympic villages from 1924 to 1948 were organised through tourist resorts in mountain resorts. To observe the first permanent model in the winter edition, we must wait until the 1952 edition in Oslo. For the 1952 edition, the Organising Committee proposed the construction of three new Olympic quarters in central areas to be reused as permanent housing for the community.

The following mega-project was justified by the need to provide housing after World War II. Therefore, the central states began to develop large projects to meet the needs induced by the increase in population. Moreover, the presence of three Olympic Villages in the territory would profoundly transform the spatial dimension of the Olympic Games. Starting with the Rome edition in 1960, the mega-event began to play an essential

Figure 11. Location of the Berlin Olympic Village 1932 (Source: Own implementation)

role in the reconfiguration and reconstruction of major European cities. The urban expansion needs induced by the population increase in Europe in the 1960s and 1970s, the profound infrastructural transformations, and the redevelopment and regeneration of urban centres will be the factors that will determine the construction of the future Olympic Villages after the Rome edition of 1960.

Therefore, it is crucial to emphasise that Rome's 1960 edition will catalyse spatial transformation through the Olympic event. Rome would become a model of urban transformation for all future candidate cities. In 1960, for the first time, the mega sports event was observed as an element to be included in the host cities' long-term strategy. Since this historic moment, the Olympic Games have catalysed more complex urban processes that transform cities over time. The interdependence between the Olympic event and the town will complexify over the century to the point of developing new forms and elements that will be collateral or secondary to the sporting event. In addition, in 1991, the IOC, through a specific amendment to the winter edition's organisation, also allowed countries without sports facilities on their territory to submit a proposal for Olympic allocation.

Figure 12. Location of the Rome 1960 Olympic Village (Source: Own implementation)

The following amendment will break down geographical boundaries and favour the proposal of new spatial models fragmented over different territories. In this way, the spatial dimension of the winter edition will grow enormously, introducing new forms of clusters of other locations connected through a trans-regional or trans-national infrastructure system. Although the IOC does not set criteria for the allocation of host cities, it can be seen in Table 18 that the summer event remains an event of the world's major metropolises. However, the winter edition has historically developed within the mountain communities, favouring the event to promote winter sports and venues for tourists.

With the election of Turin in 2006, the winter edition was transformed into a metropolitan edition that manifests itself within a regional dimension of the Olympic territory. Today, the winter edition, like the

Table 16. Population of the editions of the Summer Olympic Games (Source: Own implementation)

EDITION	POPULATION OF THE CITY
BERLIN 1936	4.242.501
HELSINKI 1952	381.000
MELBOURNE 1956	5.000.000
ROMA 1960	2.455.581
TOKYO 1964	11.829.000
MEXICO 1968	19.400.000
MUNICH 1972	2.316.000
MONTREAL 1976	2.950.000
MOSCA 1980	13.200.000
LOS ANGELES 1984	3.400.000
SEOUL 1988	10.100.000
BARCELONA 1992	1.643.000
ATLANTA 1996	394.000
SYDNEY 2000	3.610.000
ATENE 2004	772.072
BEIJING 2008	19.612.368
LONDON 2012	8.174.000
RIO 2016	13.047.000
TOKYO 2020	37.435.191
MEDIA	8.419.038

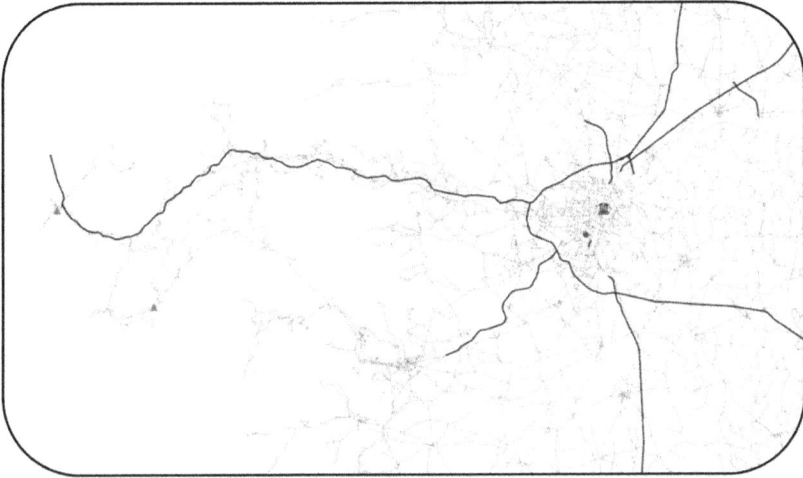

Figure 13. Olympic space in Turin 2006 (Source: Own implementation)

summer edition, is awarded to metropolitan cities that want to benefit from the Olympic event to reconfigure or expand their regional territory. Table 19 shows that the winter edition cities have an average population of at least 1.4 million. The summer cities are megacities with an average of 8.5 million inhabitants. The size of the Olympic cities, depending on the edition they host, entails and implies the execution of completely different projects for the specific site.

Therefore, the Olympic Games involve a great deal of effort and concentration of resources in a limited timeframe, which places high demands on the availability of service infrastructure and temporary accommodation (della Sala, 2022). Therefore, as Table 20 shows, most Olympic venues are located in the world's most influential megacities.

Observing the evolution of Olympic cities through urban forms of accommodation is seen as a moment of reflection on the possible evolution of modern cities over the century. Moreover, as we will observe in the following sections, the Olympic Villages induced several changes and restructuring of the urban landscape, imposing themselves as a tangible legacy of the Olympic legacy in the host cities.

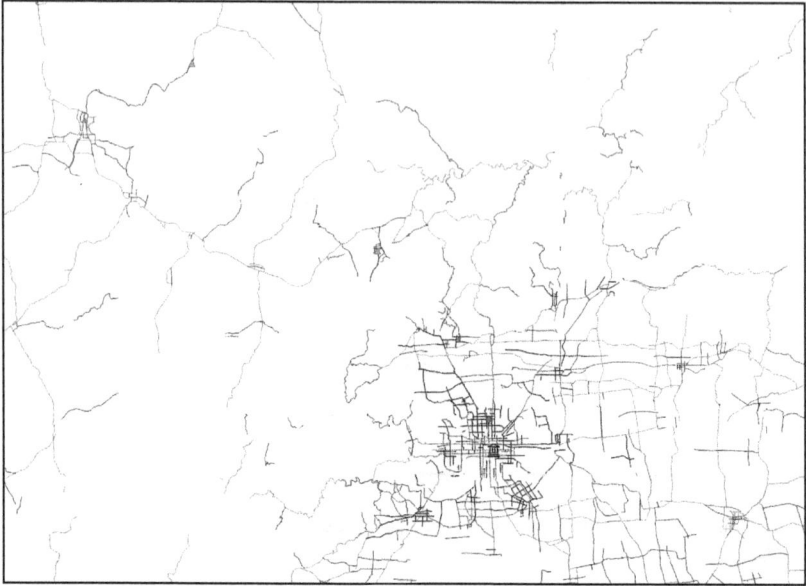

Figure 14. Olympic space in Beijing 2022 (Source: Own implementation)

4.2.2. Evolution of the location of the Olympic Villages

Before advancing the analysis regarding the different evolutionary phases and spatial patterns of the Olympic Villages, I would like to observe the evolution of the distances between the two main structures in the summer and winter editions: the Olympic Village and the Olympic Stadium. Furthermore, to keep the evolution of the spatial patterns, the study analyses the distances between the Olympic Village and the administrative centre of the summer and winter host cities. The following parameters observed in Tables 20 and 21 allow us to reflect on the evolution of the location of the Olympic Villages in the two editions disputed over time. The location of the Olympic accommodation will allow us to analyse the spatial patterns observed over time and advance new hypotheses related to the development of the Olympic Village as a central urban element of the modern Olympic event.

Table 17. Population of the editions of the Winter Olympic Games (Source: Own implementation)

EDITION	POPULATION OF THE CITY
OSLO 1952	447.000
SQUAW VALLEY 1960	4.000
INNSBRUCK 1964	100.000
GRENOBLE 1968	180.000
SAPPORO 1972	1.000.000
INNSBUCK 1976	117.000
LAKE PLACID 1980	5.000
SARAJEVO 1984	448.000
CALGARY 1988	640.000
ALBERTVILLE 1992	20.000
LILLEHAMMER 1994	23.000
NAGANO 1998	361.000
SALT LAKE 2002	174.348
TURIN 2006	900.000
VANCOUVER 2010	603.400
SOCHI 2014	364.000
PYEONGCHANG 1918	43.600
BEIJING 2022	19.638.000
MEDIA	1.392.686
MAX	19.638.000
MIN	4.000

Looking at the summer editions through the distance between the Olympic Village and the main competition venue, the Olympic Stadium was located at an average distance of 7.63 km. Meanwhile, the distance between the administrative centre and the Olympic accommodation averaged 14.13 km.

On the other hand, in the winter edition, we can observe an average distance of 16.74 km between the Olympic Village and the Olympic Stadium.

Table 18. Distance of the Olympic Village from the stadium and the Summer
Olympics administrative centre (Source: Own implementation)

EDITION	Distance from the main stadium	Distance from the administrative centre of the city
PARIS 1924	950 m	17.2 km
LOS ANGELES 1932	8.2 km	17.5 km
BERLIN 1936	21.1 km	31 km
HELSINKI 1952	2.4 km	6.9 km
MELBOURNE 1956	14.0 km	15.3 km
ROME 1960	3.1 km	6.0 km
TOKYO 1964	3.1 km	3.4 km
MEXICO 1968	4.2 km	21.5 km
MUNICH 1972	850 m	8.2 km
MONTREAL 1976	1.1 km	5.5 km
MOSCOW 1980	17.85 km (average)	18.5 km (average)
LOS ANGELES 1984	1.7 km	8.2 km
SEOUL 1988	6.5 km	18.5 km
BARCELONA 1992	7.1 km	3.0 km
ATLANTA 1996	7.8 km	5.6 km
SYDNEY 2000	2.6 km	18.9 km
ATHENS 2004	15.6 km	21.0 km
BEIJING 2008	900 m	16.0 km
LONDON 2012	2.5 km	10.1 km
RIO 2016	28.6 km	28.2 km
TOKYO 2020	9.9 km	16.3 km
MEDIA	7.63 km	14.13 km
MAX	28.6 km	31 km
MIN	0.85 km	3 km

Meanwhile, the distance between the Olympic Village and the administrative centre reached an average of 22.31 km.

Through the observation of the location of the Olympic Village, it can
be observed that, in the summer editions over time, the location of the accommodation has adopted a spatial pattern that continues to approximate
the central fabric of the host cities. The only exceptions in the twenty-first
century are Athens in 2004 and Rio in 2016, which, as we will see in the

Figure 15. Location of the Tokyo 2020 Olympic Village (Source: Own implementation)

following sections, reintroduced a peripheral model for the organisation of the Olympic event. Meanwhile, in the winter edition, the distance between the Olympic Village and the administrative centre continues to evolve, and for the Beijing edition in 2022, the average distance reached 115.63 km.

In addition, some winter editions, such as Oslo, Turin, Sochi, PyeongChang and Beijing, have arranged multiple Olympic Villages near the competition venues. The specific distances to each Olympic Village in an organisation with numerous permanent lodgings are given in Table 21.

The latest winter edition of Beijing 2022 allows us to observe a new spatial dimension of the event, which in some cases reaches a distance of 239 km from the administrative centre of the city to the Olympic Village in the mountainous areas. In conclusion, the spatial dimension in the winter edition has been transformed into a regional organisation involving considerable resources in infrastructural works for rapid connections between the various competition venues.

By adopting the following spatial model, the host city has become a promoter of a larger and more complex territory. Therefore, the next dimension of the winter edition will be characterised by a transnational space that will imply new reserves and spatial organisation models to comply with IOC standards.

Table 19. Distance of the Olympic Village from the stadium and the Winter Olympics administrative centre (Source: Own implementation)

EDITION	Distance from the main stadium	Distance from the administrative centre of the city
OSLO 1952	2.77 km (average)	4.87 km (average)
SQUAW VALLEY 1960	0.8 km	13 km
INNSBRUCK 1964	5.8 km	5.0 km
GRENOBLE 1968	600 m	4.0 km
SAPPORO 1972	1.7 km	9.1 km
INNSBRUCK 1976	5.6 km	6.4 km
LAKE PLACID 1980	10.1 km	10.4 km
SARAJEVO 1984	8.6 km	8.2 km
CALGARY 1988	1.3 km	8.6 km
ALBERTVILLE 1992	36.3 km	35.1 km
LILLEHAMMER 1994	4.3 km	3.4 km
NAGANO 1998	3.8 km	9.3 km
SALT LAKE 2002	1.6 km.	9.6 km
TURIN 2006	62.17 km (average)	65.43 km (average)
VANCOUVER 2010	1.2 km	1.5 km
SOCHI 2014	40.13 km (average)	64.33 km (average)
PYEONGCHANG 1918	13.4 km (average)	27.8 km (average)
BEIJING 2022	101.17 km (average)	115.63 km (average)
MEDIA	16.74 km	22.31 km
MAX	101.17 km	115.63 km
MIN	0.6 km	1.5 km

Table 20. Distance of Olympic Village sub-venues from the stadium and the Winter Olympic Games administrative centre (Source: Own implementation)

EDITION	Distance from the main stadium	Distance from the administrative centre of the city
OSLO 1952		
ULLEVAL	2.2 km	4.3 km
SOGN	4.3 km	6.6 km
ILLA	1.8 km	3.7 km
TURIN 2006		
MAIN VILLAGE	2.2 km	5.9 km
BARDONECCHIA	94.1 km	89.4 km
SESTRIERE	90.2 km	101 km
SOCHI 2014		
ROSA KHUTOR	58.1 km	76.6 km
MAIN VILLAGE	1.4 km	34.2 km
SLOBODA	60.9 km	82.2 km
PYEONGCHANG 2018		
Village 1	7.2 km	23.1 km
Gangneung	19.6 km	32.5 km
BEIJING 2022		
Main Village	2.0 km	17.1 km
Yanqing	72.5 km	90.8 km
Zhangjlakou	229 km	239 km

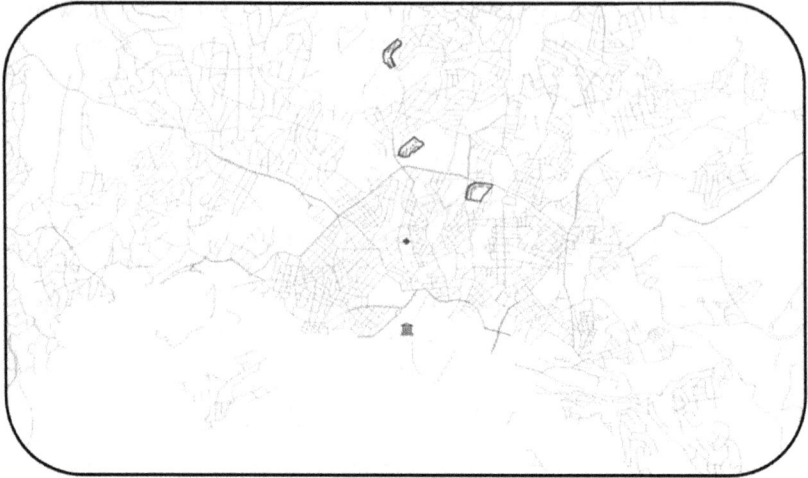

Figure 16. Spatial organisation in Oslo 1952 (Source: Own implementation)

4.2.3. Use and development of the Olympic Village area

As noted above, the Olympic Village area involves not only the construction of housing but also sports facilities, public parks, amphitheatres, services and other elements that are fundamental to the exploitation of the area in the post-event period. Therefore, the choice of the area is an essential element for allocating the work and exploiting the accommodation in the post-Olympic period. Looking at the Table 23, we can identify different uses of the area in the post-Olympic phase. Uses allow us to reflect on the transformation of urban, mountain or metropolitan spaces. Considering the use of the area before the Olympic edition, between the two editions, summer and Winter, we can observe a common language in the choice of the available areas, which will become public property after the subsequent expropriation. Regarding the use of the area in the post-event phase, in the summer edition, there is the possibility of integrating housing with offices and converting the entire project area into a public park, turning it into a mixed space. In the winter edition, on the other hand, the sports facilities and the Olympic Village are the only elements shared by the different experiences analysed.

Table 21. Use of the area of the Summer Olympic Village before and after the Olympic Games (Source: Own implementation)

City	Previous use of selected areas	Use of areas chosen after the Games
PARIS 1924	Lands	
LOS ANGELES 1932	Lands	
BERLIN 1936	Land. Forests	Olympic Village, Olympic Park, sports facilities
HELSINKI 1952	Lands	Olympic Village, sports facilities
MELBOURNE 1956	Lands	Olympic Village, sports facilities
ROME 1960	Stadium, land, barracks	Olympic Village, sports facilities
TOKYO 1964	Military area	Public Park
MEXICO 1968	Lands	Olympic Village, sports facilities
MUNICH 1972	Land, forest	Olympic Park, Olympic Village, transport, sports facilities
MONTREAL 1976	Lands	Olympic Park, Olympic Village
MOSCOW 1980	Lands	Olympic Villages
LOS ANGELES 1984	University area	New student accommodation
SEOUL 1988	Contaminated, unhealthy sites	Olympic Park, sports facilities, Olympic Village, healthy spaces
BARCELONA 1992	Brownfields, land	Olympic Village, residential area, services, sports facilities, harbour
ATLANTA 1996	Residential area in the city centre	Office space, new student accommodation, sports facilities
SYDNEY 2000	Land, abandoned spaces	Olympic Park, residential area, park, Olympic Village
ATHENS 2004	Military and Industrial Area	Sports facilities, port, Olympic Village
BEIJING 2008	Underdeveloped area	Olympic Village, park, sports facilities
LONDON 2012	Industrial area	Olympic Village, offices, Olympic Park
RIO 2016	Land, barracks	Olympic Park, Olympic Villages
TOKYO 2020	Land and residential space in the city centre	Olympic Villages

Table 22. Use of the area of the Winter Olympic Village before and after the Olympic Games (Source: Own implementation)

City	Previous use of selected areas	Use of selected areas after the Games
OSLO 1952	Earth	Olympic Village, sports facilities
SQUAW VALLEY 1960	Land, hotels	Sports facilities, hotels
INNSBRUCK 1964	Earth	Olympic Village, sports facilities
GRENOBLE 1968	Earth	Olympic Village, sports facilities
SAPPORO 1972	Land	Olympic Village, sports facilities
INNSBRUCK 1976	Land	Olympic Village
LAKE PLACID 1980	Military area	Prison
SARAJEVO 1984	Lands	Sports facilities, hotels
CALGARY 1988	University area	New student accommodation
ALBERTVILLE 1992	Hotels	New tourist accommodation, sports facilities
LILLEHAMMER 1994	Land, forest	Sports facilities
NAGANO 1998	Lands	Olympic Village, sports facilities
SALT LAKE 2002	Military area	New student accommodation
TURIN 2006	Brownfields	Tourist accommodation, Olympic Village, sports facilities
VANCOUVER 2010	University area	Olympic Village
SOCHI 2014	Land, forest	Olympic Village
PYEONGCHANG 1918	Lands	Sports facilities, Olympic Villages
BEIJING 2022	Lands	Olympic villages, sports facilities

The Evolution of the Olympic Villages

ABSTRACT

After observing the evolution of the Olympic Villages throughout history and introducing the spatial models, the following chapter will analyse the Olympic accommodation through the cartographic representations advanced by della Sala (2022). The different evolutionary phases of the Olympic Villages in the summer and winter editions will be introduced by analysing the host cities' various forms and organisation models. Finally, the chapter will allow us to see the other formal languages used over time in the two Olympic editions. The different spatial models, the evolution of the location of the Olympic Village and the formal language will help us observe the different urban strategies provided by each host country throughout history.

5.1. The different stages in the evolution of the Summer Olympic Village in the urban fabric of the metropolis

5.1.1 Phase 1: Temporary accommodation in military sites (1896–1920)

From the first edition of the Olympic event, held in Athens 1896, until Paris 1924, the event was organised to coincide with the Universal Expositions, guaranteeing high internal visibility. Therefore, the following editions were characterised by the self-organisation of the host countries and the use of venues that were not strictly sports venues. Subsequently, as the number of athletes increased, hotels, military sites, ships and other places that could temporarily accommodate beds were chosen as accommodation.

Figure 17. Evolution of the Summer Olympic Village (Source: Own implementation from OCOG, 1924, 1932, 1960, 1984, 1992, 2000, 2012, 2024, 2028)

Los Angeles 1932

Figure 18. Spatial organisation in Los Angeles 1932 (Source: Own implementation)

5.1.2 Phase 2: Demountable housing in peripheral areas (1924–1932)

Only after the Paris edition of 1924 did the Olympic Village become a fundamental element in the design and development of the Olympic event. In this second phase, the Olympic accommodations were arranged through temporary solutions involving modular and removable materials. The first Olympic Village in Paris was designed close to the main stadium by providing temporary wooden huts and integrating communal facilities. The Paris experience was the first to enjoy essential services such as a post office, a leisure area, and primary services. Services were to be converted into constituent elements of the international area of the Olympic Village. Later, with the Los Angeles edition in 1932, the Olympic Village will adopt a new form that will introduce a new development model for subsequent Olympic cities. Providing housing in a suburban area will stimulate the transformation of neighbourhoods and sports facilities in other locations worldwide. In addition, the location of the Olympic Village will promote temporary urbanism by introducing inexpensive and demountable materials. The Los Angeles Village was the first to include an amphitheatre, hospital, church and other training facilities for Olympic athletes. In this way, the practice of the old Olympic Games, in which athletes were temporarily housed in the community of Elis (OCOG, 1933), was re-established. The Village of 1932 is based on the repetition of a prefabricated model around a morphology similar to the circus of ancient Rome. The 550 wooden huts will be the promoters of Olympic accommodation throughout the centuries[1].

5.1.3 Phase 3: Construction of a sports district in peripheral areas (1936–1956)

The inspiring design of Los Angeles in 1932 served as an inspiring model for the Berlin Organising Committee, which in 1936 introduced the construction of a new sports quarter in a suburban area of the city. The Berlin edition would be the first to offer a permanent solution for Olympic accommodation, introducing new elements for later reuse as residential accommodation in the post-Olympic phase. The entire military area of Döberitz will include the sports facilities, reception, restaurants, conference rooms and services included in previous editions. Thus, the 1936

---

1   In the Los Angeles 1932 edition, the candidate cities will be obliged to propose an Olympic site to celebrate the Olympic medals.

project introduced a new concept of Olympic accommodation that would be inspirational throughout the twentieth century. From now on, the accommodation site will be promoted as a multifunctional venue for future host cities. The Berlin model is based on applying the satellite city or garden city model in a peripheral space. It introduces a new model of Olympic urbanism characterised by the combination of sports facilities and Olympic accommodation. However, after the initial premises, in the post-Olympic period, the space was only used as a military school and, to this day, the complex is abandoned and without a definite destination.

The Berlin edition would inspire future projects in Helsinki in 1952 and Melbourne in 1956. In the next phase, the Olympic cities planned to build a new sports district in suburban areas, focusing on constructing sports facilities and housing. Helsinki's 1952 project was the first model of residential reuse in the post-Olympic phase. Later, in 1956, Melbourne proposed the construction of a new neighbourhood in an abandoned area. However, Helsinki will be regarded as the first post-World War II Olympic village to introduce a transformation in the accommodation of athletes. The provision of various accommodations for athletes participating in rowing competitions implied the establishment of several Olympic villages in the area. Therefore, from now on, participants in sailing and rowing competitions will be accommodated in temporary facilities close to the competition venues. The Olympic Village in Helsinki housed approximately 4,800 people in thirteen four-storey buildings (COJO, 1953).

The Olympic Quarter included a restaurant, a cinema, saunas and toilets. Meanwhile, for the 1956 Melbourne edition, the model adopted by the Organising Committee was centred on a derelict suburb (Heidelberg) 15.3 km from Melbourne's administrative centre. Heidelberg was built at a distance of 14 km from the Olympic Stadium and, for the first time in Olympic history, men and women were housed in the same Olympic Village. Melbourne's plan was to build 365 homes, which were sold to residents in the post-event phase (COJO, 1958).

Berlín 1936

Figure 19 Spatial organisation in Berlin 1936 (Source: Own implementation)

5.1.4 Phase 4: Modern housing: Rationalism and functionalism (1960–1988)

Continuing the evolutionary observation of Olympic accommodation, the next stage, the 1960 Rome edition, will be the promoter of a new design scale that will inevitably influence the scale of subsequent projects. The Rome project is recognised as one of the most ambitious. In addition to being a permanent legacy for the city of Rome, the *Foro Italico* park will become one of the first models of a sports city designed by the leading Italian rationalist architects of the 1930s. The entire area of the Olympic Village covers an area of 0.30 km². The residential complex was created following the guidelines of the rationalist style and, in the period following the event, was converted into a district for civil servants (COJO, 1963). Undoubtedly, the Italian edition was a source of inspiration for promoting a modernist style within future candidate cities. However, in the 1964 Tokyo edition, the transformation project included infrastructural development as the main objective for the post-event phase (OCOG, 1966). Therefore, the Olympic Village was planned in a peripheral area of the city through temporary accommodation made of inexpensive and demountable materials[2]. Subsequently, the Olympic Village area will be transformed into one of the most famous parks in Japan, the *Yoyogi* Park[3]. The housing, dismantled after the event, will be the only exception regarding major works and significant investments in construction. The Mexico 1968 edition will adopt a satellite city planning philosophy in a peripheral space of the metropolis, proposing temporary housing solutions to be used as residential accommodation in the post-event phase. The construction of a new sports district with the aid of flat block buildings introduced a new philosophy and a new height never before achieved. The entire Olympic Village area included the construction of two different villas, one for athletes and journalists called "Miguel Hidalgo" and one for Olympic judges and volunteers known as "Narciso Mendoza". The central Village was built near the main stadium with twenty-nine buildings between six and ten storeys (COJO, 1969). The Judges' and Volunteers' Village comprised 686 two-storey and 90 four-storey buildings (OCOG, 1969). The Housing Project 1968 can be defined as the construction of a self-sufficient residential city that would house almost 10,000 people in the post-Olympic period.

As stated earlier, both Olympic Villages were intended to be sold to middle-class families in the post-Olympic period. During the Mexico edition, the IOC published guidelines for designing and planning the Olympic Village for the first time.

Introducing essential requirements, consideration of access, and creating middle-class facilities were some themes that promoted a new type of centralised development for Olympic athletes' accommodation. In the next edition in Munich in 1972, the Organising Committee proposed the development of a large sports park of 473,000 m² in a central area. The area was destined to become one of the largest sports parks in the world

2   Before the event, the area was owned by the US military. The location was expropriated, and ownership was transferred to the Japanese state.

3   After the Olympic event, the park has become one of the symbolic spaces of the transformation of the city of Tokyo. Today, only one of the houses built for the Olympic event can be found in the park.

(COJO, 1974). In the next phase, the vast housing complex was converted into residential accommodation for young couples, and new buildings for university accommodation were introduced. In contrast to the previous complexes, Munich's housing will be used as integration elements within the central spaces of the host city. Thus, Munich will define a new philosophy of urban growth. Munich's 1972 and Montreal's 1976 housing will undoubtedly be two initiatives that will reinforce the creation of artefacts through the use of a brutalist construction philosophy. Both Olympic areas represent a unique urban phenomenon. The Olympic Village and sports facilities are embedded within a large sports area that only needs to expand or reconfigure after the post-event phase. The Munich complex and the Olympic Park were located 8.2 km from the city centre. In addition, the large complex included a set of buildings that could hold more than 12,000 people. Therefore, the concept of a residential and international area was introduced for the first time in 1972. From this moment on, the planning of Olympic accommodation will focus on the functionality of the buildings while considering leisure and green spaces.

Munich 1972

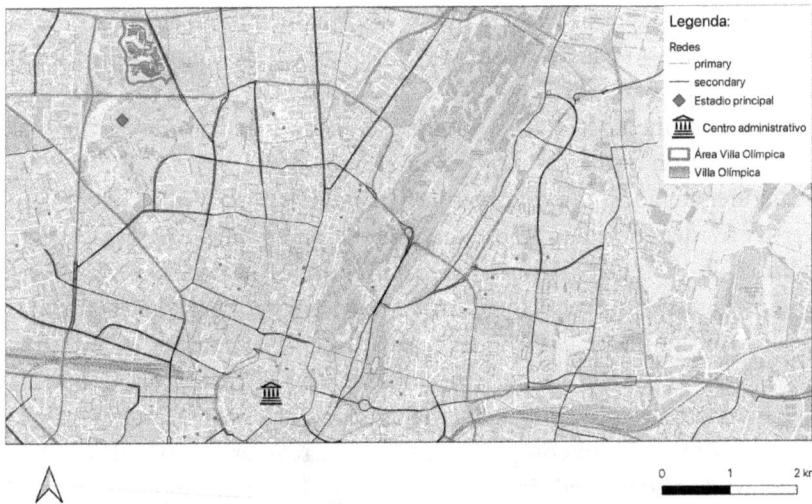

Figure 20. Spatial organisation in Munich 1972 (Source: Own implementation)

Subsequently, the Montreal edition will follow the same philosophy as Munich, building a large housing complex near the sports facilities turning the district into a huge sports centre for future events. However, the plan to make the Olympic Village a vast green area provoked some protests from local citizens. Therefore, the Montreal edition will represent a significant crisis for the Olympic movement, forcing it to rethink the allocation of future editions to not maintain the promotion of sport worldwide. The Montreal Olympic Village consists of four 19-storey blocks that include services and offices. After the event, the accommodation was sold to families (COJO, 1978).

Meanwhile, the 1980 Moscow edition was part of a twenty-year (1971–1990) metropolitan reconstruction and development project in Moscow. The plan included reorganising the urban fabric and installing new sports infrastructure (OCOG, 1981). In addition, the plan envisaged a single development strategy in eight different areas that were given their own centre by building satellite areas in the metropolitan dimension. The organising committee proposed the construction of Olympic Villages in two regions included in the infrastructure reform of the Russian capital. Therefore, the accommodation construction would only result from a defined territorial expansion within the twenty-year project. Both Olympic Villages comprised thirty-four residential blocks of sixteen to eighteen storeys in Soviet Brutalist style. In the post-event phase, all housing was intended for young families (OCOG, 1981).

Meanwhile, the 1984 Los Angeles edition is recognised as the first private edition of the Olympic event and the first edition to use existing structures. In addition, the provision of the Olympic Village at the university residences will promote a new development philosophy for future editions. Los Angeles has been an inspiration for using existing facilities and university residences to advance knowledge through the venues and services of significant universities. The University of Southern California, the University of California at Los Angeles, and the University of California at Santa Barbara could host more than 12,000 participants, offering all the necessary facilities to comply with Olympic regulations (COJO, 1985). Meanwhile, the 1988 edition in Seoul followed the philosophy of the previous editions, proposing a comprehensive reconstruction project that would include Olympic housing at its core. The Seoul Olympic Village will be built in a suburban area near the central stadium and competition venues. The block housing will be converted into permanent residences and available to citizens in the post-event phase. The Seoul Olympic Village was a large housing project comprising eighty-six blocks of twenty-six-storey buildings occupying a total area of 1.19 km² (COJO, 1989). In addition, the housing project was included in the metropolitan city's strategic plan to improve services and health conditions for its citizens. The entire Olympic area housed more than 14,000 people.

5.1.5 Phase 5: The promotion of housing as a tool for a new lifestyle (1992–2004)

The industrial crisis of the late 1980s strongly conditioned the development of the world's major industrial cities. However, Barcelona's candidature in 1992 initiated a new design and development phase for the Olympic Games. The regeneration and transformation of enormous abandoned and disused land into central areas of the city will significantly condition the location and philosophy of the Olympic venues. The transformation of the post-industrial city through mixed financing opens the door to new economic flows related to tourism and services (Venturi, 1994). The Olympic Village in Barcelona is at the centre of one of the Catalan capital's main objectives: to break down borders and open the city to the sea. Therefore, the housing is in a privileged location, close to the waterfront and the Olympic harbour. The Barcelona housing project will allow the Olympic Village to be a catalyst for transforming the entire industrial area. In addition, the central location of the housing will make it possible to observe how residences can be used to promote a new lifestyle in host cities.

The Olympic Village in the Poble Nou district, a former industrial area, is seen as the starting point for the development of the organising committee's objectives: to re-establish the connection between the city and the sea, to continue with the urban model of Cerdá and to introduce new uses for the spaces (COJO, 1992). The Olympic Village comprises eighteen flat blocks ranging from two to nine storeys in height. The project integrated creating large spaces specifically for commercial services, hotels, and offices. The Barcelona housing project can be seen as inspiring planning for housing construction through mixed financing in an area with a high risk of speculation in the post-event phase. The introduction of diverse strategies in Barcelona will ensure a new promotion of Olympic housing, implying the creation of a new international image for the Catalan capital. One Olympic cycle later, the 1996 Atlanta edition will propose a temporary solution by using university residences and tensile structures to provide services. Atlanta's organisational model was inspired by the one offered by Los Angeles in 1984; the Olympic Village was located on the Georgia Institute of Technology campus due to its proximity to the sports facilities. Meanwhile, the use of existing infrastructure was complemented by the construction of two new buildings that increased the availability of university accommodation in the city after the Olympic event (COJO, 1996).

Subsequently, in 2000, Sydney will be the first city in the world to include sustainable solutions for the construction and maintenance of the Olympic Village. Sydney's plan called for constructing a new sports district on a brownfield site near the Olympic Park, which was developed to promote sport through a large sports area (Davidson, 2012).

The Sydney Olympic Village was designed by providing 870 flats converted into residences in the post-event phase, forming a new neighbourhood and promoting the expansion of the Australian metropolis (Blunden, 2012). The entire neighbourhood covers an area of 510,861 m². Furthermore, the use of sustainable materials, the provision of solar panels, and water recycling paved the way for a new model of sustainable housing that will be adopted by other cities in the future (Spooner, 2000). However, the Sydney project will implement the number of dwellings built, introducing new design elements for future editions.

Meanwhile, in the next edition in Athens 2004, the Olympic Village was built in a suburban area of 1.09 km². The entire area was transformed into a large sports park where most sports facilities were located, except the Olympic Stadium and other competition sites. By analogy, the Olympic Village in Athens can be identified as a satellite city project at a distance of 21 km from the city's administrative centre (della Sala, 2022). The later

Barcelona 1992

Figure 21. Spatial organisation in Barcelona 1992 (Source: Own implementation)

project was intended to provide new residential accommodation in the post-Olympic phase. The Olympic Village consists of 366 four-storey blocks (COJO, 2005). However, in the post-Olympic period, Greece, hit by the severe economic crisis, abandoned the project, occupied by immigrants living in the Olympic area today. This mega-project marked a new situation for Olympic housing, leading once again to a redefinition and rethinking of the size of the projects. Therefore, the Athens project will go down in history as one of the worst results observables in the history of Olympic Villages.

5.1.6 Phase 6: Sustainability and heritage as a stimulus for metropolitan development (2008–2028)

In the next phase, the Beijing edition will promote a further change for the future of Olympic accommodation. Choosing a green area, using sustainable materials and environmentally friendly solutions, will emphasise environmental protection and sustainable development (Smith, 2007). The Beijing project was included in an area north of the city promoted by the 1990 Asian Games, and the city wanted to turn into a huge sports park (COJO, 2010).

Therefore, the housing construction will be part of a significant housing strategy reserved for Communist Party members. The accommodations were intended to be public residences in the post-Olympic period. However, Beijing's project will not only advance a new model of housing construction but will also promote a significant expulsion of citizens and the destruction of existing buildings. It is estimated that more than two million people have been excluded from their homes and neighbourhoods to make way for the new Olympic Village. Subsequently, the 2012 London project will be seen as another milestone in the evolution of a new model of metropolitan housing in vacant areas with high development potential (Poynter, 2012). The East London area has been included in an infrastructural reorganisation project and an expansion vision for the city (Smith, 2014). However, the London 2012 project will be a crucial moment to observe privately financed housing development in an area that, in the post-Olympic period, is subject to a high risk of property speculation. Therefore, the London Olympic Village will be a watershed moment for the host city's post-Olympic planning and housing legacy. Redeveloping derelict spaces and introducing new environmental protection measures will allow the British capital to own a large sports park in a central area. In addition, including services and offices in the area will promote a new type of mixed housing that will favour the diversification of residences for more excellent optimisation in the post-Olympic phase.

The 2016 edition of Rio de Janeiro marked another crisis in the history of Olympic housing. The housing project was organised by defining a new suburban area that could serve as a satellite city. Before becoming a new sports district, the entire area was occupied by shacks and unconventional structures, resulting in citizens moving inland (della Sala, 2022). The Rio Olympic Village was planned in an area 28.2 km away from the city centre. In addition, the project was included in an infrastructure reform and the reorganisation of the new rapid transit service. This new service will promote new modes of transport to connect the Olympic areas. The

Londres 2012

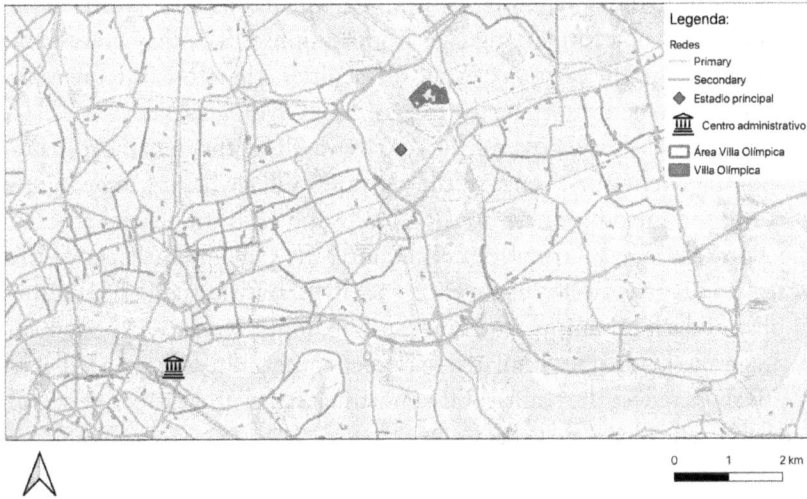

Figure 22. Spatial organisation in London 2012 (Source: Own implementation)

Olympic Village has thirty-one flat blocks on seventeen floors (COJO, 2010). A sizeable urban planning project intended to be converted into a new, accessible neighbourhood for all citizens. However, the Olympic Village in Rio, like the one in Athens, can be considered two failed satellite city models. Models that were abandoned and sometimes illegally occupied in the post-Olympic period.

Finally, Tokyo 2020 developed a similar formal language to that observed in London in 2012. The area was chosen due to its strategic importance for the redefinition and redevelopment of the island hosting international events. As in London, the construction of new accommodation took place through a private agreement in a central city area, guaranteeing its reuse in the post-event phase. However, the Olympic Village project envisaged the subsequent construction of two skyscrapers to accommodate offices and housing (COJO, 2013). The 2020 Tokyo edition will confirm the importance of the Olympic Village as a tool for redeveloping empty spaces in metropolitan cities and as a catalyst for new socio-economic transformations.

In conclusion, the assignments of Paris 2024 and Los Angeles 2028 will allow us to observe new housing models, including new construction elements and a new formal language. Both editions will be characterised by temporality and the existence of sports facilities. Therefore, the Olympic Village's construction will be the focus of the Olympic project.

We have observed how, during the twentieth century and more markedly in the twenty-first century, the Olympic Village has become an opportunity to introduce new housing philosophies in the central areas of the world's largest metropolises. Therefore, the Olympic Villages of the next decade of the twenty-first century clearly show the main orientations of postmodern urbanism, including sustainability, security, experiences, heritage and landscape. Similarly, the experiences of the Olympic Villages at this stage reflect the reality of the most successful urbanism in terms of economic efficiency (della Sala, 2022).

Tokio 2020

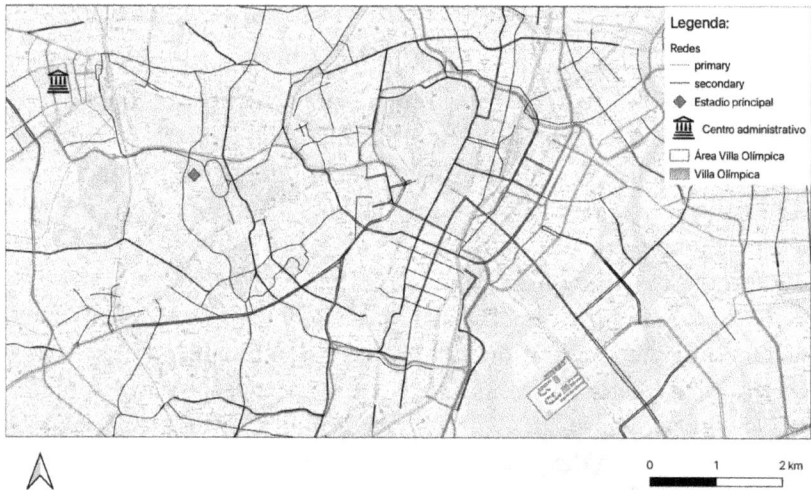

Figure 23. Spatial organisation in Tokyo 2020 (Source: Own implementation)

Table 23. Stages of the Olympic Villages at the Summer Olympics (Source: Own implementation)

Phase I	1896-1920	Temporary accommodation	Prospects for the development of an Olympic Village Use of hotels and military spaces
Phase II	1924-1932	Peripheral areas Removable housings	Prospects for the development of a permanent Olympic Village Specific area for holding the event Development of a temporary location for Olympic accommodation
Phase III	1936-1956	Establishment of a sports district Permanent housing	Creation of a sports quarter in the peripheral areas of the cities Sports facilities and services The foundations are laid for developing residential accommodation in the post-Olympic phase.
Phase IV	1960-1988	Expansion of the Olympic Village Residential development tool	Increase in the number of Olympic athletes. Public sector funding for the construction of new accommodations Increasing the size of the Olympic Village area
Phase V	1992-2004	Olympic Village in the city The stimulus for the transformation of abandoned areas	Olympic Village as part of the revaluation of industrial areas Mixed economy for the construction of the residences The Olympic Village is a tool for the promotion of a new lifestyle.

(*continued*)

Table 23. Continued

Phase VI	2008-2028	Transformation of the Olympic Village Global cities	Metropolitan development in empty spaces. Tool for the redefinition and reorganisation of the economy of the Olympic area. Greater emphasis on the protection of the environment and the sustainable development Olympic legacy assumes great importance in planning for post-Olympic phase Inclusion of services and offices Mixed housing solutions for social inclusion in cities

5.1.1. Spatial models of the Olympic Summer Villages

The following projects reflect the major urban transformations of the main Summer Olympic cities in the urban fabric of the host cities. Historically, the Summer Olympic Villages have been integrated into the interventionist policies of cities through different urban models about the requirements of each host city. Each Olympic Village has different configurations and constraints that reflect the housing philosophies of every town. In general, the rationalist model through block housing represents one of the most used models for the strategic development of the Olympic Village. Meanwhile, the decentralised model is observed in cities that had or needed a tremendous infrastructural transformation to connect different areas in the metropolitan territory. Finally, urban planning models in central locations have, since London 2012, established themselves as the most used by those large metropolises that want to reconvert significant areas. Areas are strategically located for the operation of new services and infrastructures (della Sala, 2022). Finally, the peripheral cluster development model in the summer edition was only used to meet the organisational requirements imposed by the IOC.

Ꝥ OLYMPIC VILLAGE
— NETWORK

MONOCENTRIC

Munich 72; Montreal 76; Atlanta 96; London 12;
Tokyo 20; Paris 24 ; Los Angeles 28

POLYCENTRIC

Moscow 80; Beijing 08

SATELLITES

Helsinki 52; Tokyo 64

CLUSTER

Melbourne 56; Roma 60;
Barcelona 92

PERIPHERY

Paris 24; Los Angeles 32;Berlin 36;
Mexico 68; Seoul 88; Sydney 00; Athens 04; Rio 16

Figure 24. Spatial models of the Olympic Village at the Summer Olympics
(Source: Own implementation)

Analysing the Olympic Villages within the transformations of Olympic cities can help observe the evolution and sensitivity of the housing issue in large contemporary metropolises. Today, the Olympic City has become a more complex geographical concept that must meet new requirements regarding air quality, water reuse, waste collection, public spaces and well-being. Therefore, Olympic cities must be seen as an open and dynamic space to reinterpret and implement new theories of sustainable development for the future of our communities. As we have observed, the Olympic Village does not have a blueprint or a reference model. However, some models have become a reference for the Olympic urbanisation of future candidates and host cities. Figure 24 shows the spatial models adopted in the summer edition.

5.2. The different stages of the development of the Winter Village on the regional territory

5.2.1 Phase 1: Promotion of mountain tourism in resorts (1924–1948)

Since the first Winter Olympic Games in Chamonix in 1924, the Olympic event has been organised in mountain areas with ski resorts. In addition, the mountain venues were equipped with hotels, resorts, a sports pavilion and different accommodation facilities. Therefore, as noted above, the winter edition up to Oslo 1952 was planned in places with a strong tourist vocation for winter sports. The allocation took place considering the presence of accommodation facilities in the area and the new plans for rustic expansion in the mountainous locations. Therefore, until Oslo 1952, the winter editions did not include constructing new permanent accommodation. In this phase, only punctual interventions were observed for building a sports pavilion or expanding accommodation. Regarding the development, Oslo in 1952 was considered the first Winter Edition city to have a permanent Olympic village (Delorme, 2014).

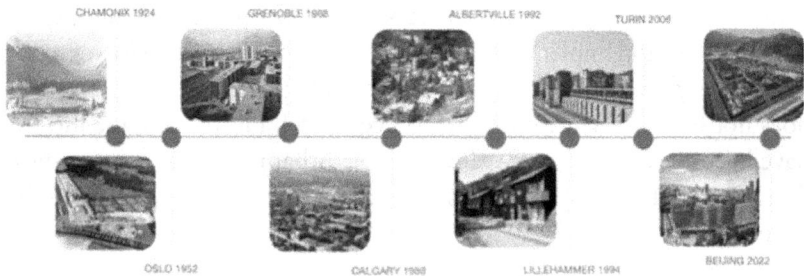

Figure 25. Evolution of the Winter Olympic Village (Source: Own implementation from COJO, 1924, 1952, 1968, 1988, 1992, 1994, 2006, 2022)

5.2.2 *Phase 2: Development of a public housing policy (1952–1964)*

The 1952 Oslo project was conceived through a polycentric spatial organisation that included the construction of three Olympic Villages in the urban fabric of the Norwegian capital. Consequently, from 1952 onwards, the Winter Olympic Village evolved into a spatial transformation model very similar to what was observed in the summer edition. For the 1952 winter event in Oslo, the athletes' accommodation was organised in three different areas of the city, integrated within the urban transformation plan (Illa, Sogn and Ulleval). Therefore, from the beginning, the Olympic Villages were intended to be self-sufficient and transform into new residential accommodations in the post-event phase. All Villages were planned in an area that included most primary and secondary services. The buildings were organised in eighteen blocks ranging in height from two to eight storeys (COJO, 1953). In this second phase, we observe a growing interest on the part of the candidate cities to host the winter edition. Therefore, the promotion of winter sports will be a new tool that cities will use to incorporate the Olympic event within the ordinary transformations of the territory[4]. Subsequently, the 1964 edition of Innsbruck proposed the construction of a new district in an area of the city included in a development plan owned by the central state. The Village was planned through the construction of four 10-storey blocks (COJO, 1967). From the planning stage, the Olympic Village was destined to become a large residential area in the post-event phase. In conclusion, in this evolutionary phase, we can observe an increase in public funding for constructing new residential housing stimulated by the growing interest in the Olympic bid.

4   With a resident population of 447,200, the city was by far the most significant centre to have hosted the Games to that date. The larger population created new opportunities for the type of facilities offered, as viability and future use after the Games were more than assured (Essex, 2017).

Oslo 1952

Figure 26. Spatial organisation in Oslo 1952 (Source: Own implementation)

5.2.3 Phase 3: Mixed housing in a regional development dimension (1968–1988)

With the consequent increase in size and interest in winter sports, cities will begin to promote a new spatial model in a larger territory. As we will observe during the Grenoble Organisation 1968, the Olympic territory will expand throughout the region, implying a rethinking of the road infrastructure. The growing demand for infrastructure in the 1960s allowed Grenoble to complete a new regional transformation project. The central government supported Grenoble's mega-project to increase tourism and trade in mountain areas. The Grenoble edition will promote new accommodation solutions in the significant fabric of the city, close to the main stadium and with the help of removable structures in the post-event phase. The structures and architectural style of the Olympic Village reflected the rationalist style that was respected.

Le Corbusier's idea for the development of new functional cities. Therefore, the Olympic quarter and residences were intended to be utilised through a mixed management that would benefit the university, citizens and tourism. The Olympic Village was organised in eleven blocks of between four and five storeys in height. In addition, creating new highways, roads, airports and railway lines would determine a new model for organising the Olympic event.

The Olympic Village was included within a priority urbanisation area, identified in the master plan as an area of new interest for regional development. Therefore, Grenoble will be the winter event that will be transformed as a regional development tool able to catalyse major infrastructural interventions. Like Grenoble, Sapporo 1972 implemented a project to redevelop and redefine urban areas and regional infrastructure (Kagaya, 1991).

Sapporo will be the first city with more than 1 million inhabitants to host the Olympic event. The Olympic Village was included in a housing development plan, which, through the provision of block structures, was transformed into residential housing in the post-Olympic phase. In the following phases, we will observe how the housing emergency is an element

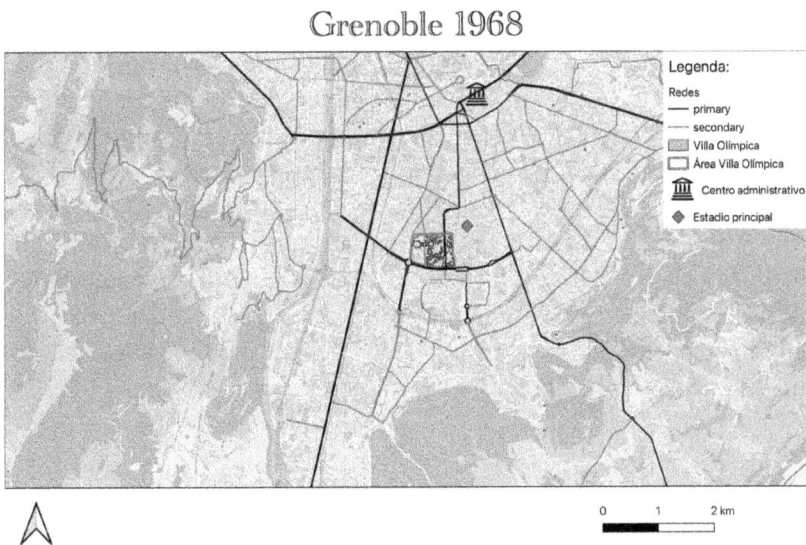

Figure 27. Spatial organisation in Grenoble 1968 (Source: Own implementation)

that will influence the layout and construction of residential projects in the candidate cities. In the meantime, only a few hotels will be built in the mountain villages for participants in Alpine sports. As observed in Grenoble, the Sapporo Olympic Village will promote a new construction model never before achieved in a winter edition. The complex was realised by constructing twenty residential blocks between five and eleven storeys high (COJO, 1973). The city of Sapporo was the first to lack the sports facilities for the event, which until Grenoble had been essential for hosting Olympic competitions. Thus, at this stage, we will observe how sports facilities became another crucial element for promoting winter sports within metropolises. Subsequently, the 1976 Innsbruck edition proposed the construction of a new Olympic Village in an area adjacent to the one built for the 1964 edition (OCOG, 1976). Knowing that the accommodation was occupied by the community, the organising committee had to provide a new housing solution that involved the same reuse philosophy for the post-Olympic period. Therefore, after the Olympic event, the area was to be transformed into a new residential neighbourhood for citizens, promoting an expansion of the neighbourhood built for the 1964 event (della Sala, 2022). Until Calgary 1988, the number of athletes steadily increased. As a result, Olympic cities began to promote new solutions for Olympic accommodation projects. Calgary is recognised as the first winter edition city to propose university-type housing solutions (Olds, 1998). In addition, the organising committee focused the project on constructing new sports facilities for university students and promoting winter sports (COJO, 1988). The Calgary solution greatly stimulated future editions, offering a new model for transforming the Winter Olympic Games. Meanwhile, in the third phase, the size of the Olympic space will increase significantly. Moreover, the increase in the number of competitions and athletes will imply the construction of new sports facilities and new housing solutions through the planning of multiple Olympic villages.

5.2.4 Phase 4: Tourism development tool (1992–2002)

Albertville 1992 is recognised as the first project that proposed multiple accommodation solutions in different mountain resorts and the main Olympic village in an area included in the tourism development of the entire region. However, Albertville 1992 proposed a polycentric spatial model that included and strengthened the region's position as an international tourist centre (Terret, 2008). The significant investment in the event allowed for the construction of new accommodation and hotel facilities throughout the Olympic area, renewing the mountain's entire infrastructure system to reposition the resorts in the infrastructure system (della Sala, 2022). The availability of eight hotel facilities to meet the requirements of the Olympic event created a new spatial dimension never seen before. However, after the Albertville edition, the IOC was concerned about the event's size and the athletes' dispersion to different mountain locations. Therefore, with the constant increase in athletes, cities will be forced to plan multiple Olympic villages. After Albertville, the Olympic event will become a new tool for rebuilding new territories and repositioning towns in the winter tourism market[5].

As early as 1994 in Lillehammer, the organisers again introduced the topic of temporary accommodation facilities for athletes. In addition, the Norwegian Organising Committee introduced the theme of sustainability and sustainable development by providing 185 removable wooden huts (COJO, 1995). The solution adopted by Lillehammer served as an inspiration for the entire Olympic movement and future cities. The Olympic Village was intended to be completely dismantled after the event. Therefore, starting with Lillehammer, the theme of sustainability and respect for the environment became essential for the winter edition. However, significant infrastructural changes and the increasing scale of the event compromised the candidate cities' environment and regional development (Spilling,

---

5   The number of overnight stays increased from 100,000 in 1989 to 700,000 in 1995. Thus, in 1996, Brides' financial situation aligned with expectations. The municipality's budget increased from 15 million francs in 1992 to 25 million francs in 1996 (Sordet, 1996).

Albertville 1992

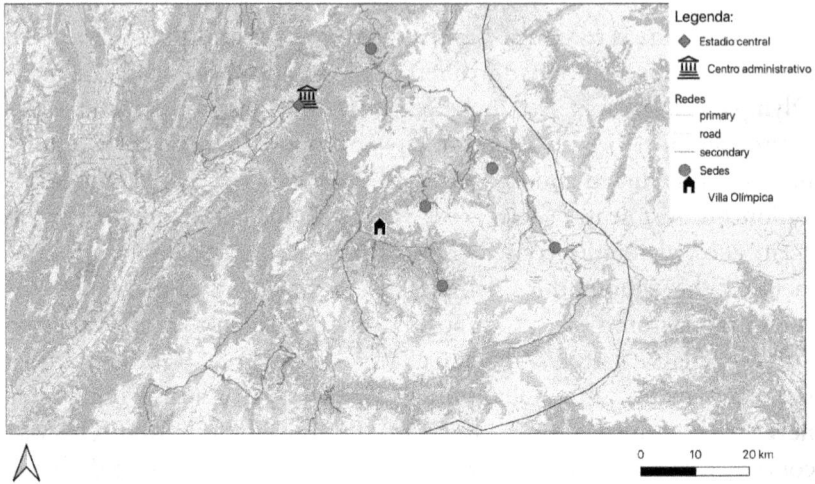

Figure 28. Spatial organisation in Albertville 1992 (Source: Own implementation)

1996). Lillehammer allowed the IOC to add sustainability as the third pillar of the Olympic movement.

Meanwhile, in the next edition, Nagano 1998 again proposed a project involving the construction of a new district and new neighbouring sports facilities. The following Japanese edition was part of a series of regional transformations that included Nagano within the new regional economy. For example, the construction of the railway line between Nagano and Tokyo would enormously change the city's economy. The Nagano Olympic Village was planned in a suburban area of the city to be converted into private residences during the post-Olympic period. The Village was realised by constructing twenty-three residential blocks ranging from two to four high storeys (COJO, 1999). Subsequently, Salt Lake in 2002, like the 1988 Calgary edition, proposed university-type housing solutions. However, the organisers developed new measures to protect the environment, including new sustainable development processes for the event (OCOG, 2002). The Salt Lake project enabled the organising committee to achieve zero carbon emissions, recognising it as one of the most sustainable events ever.

5.2.5   Phase 5: Multiple Olympic Villages in a regionalisation context (2006–2022)

Subsequently, in the fifth phase, the Turin 2006 edition again introduced a new spatial dimension to the Olympic project. The Turin edition would include and adopt sustainable development practices by applying a strategic environmental development assessment throughout the Olympic process. The organisers proposed a new spatial model with the metropolitan city as the venue for the ceremonies, ice sports competitions and the central Olympic Village. On the other hand, the project proposed a spatial organisation in two mountain resorts: Bardonecchia and Sestriere. The arrangement of the three Olympic villages and the spatial dimension of Turin transformed the area permanently, favouring the increase of winter tourism and the economic attractiveness of the Savoy capital. The Turin Olympic Village was designed in a disused area included in the city of Turin's development plan. From the beginning, the organising committee had the objective of reconverting the entire area with mixed functions: services, residences, shops and offices. However, the Turin Olympic Village has never fully become a diverse area, as it suffers from many structural problems and was occupied by people awaiting political asylum in 2012. Whereas the Olympic Villages in mountain resorts were intended from the planning stage to be converted into hotel accommodations and family holiday flats. The renovation of the Bardonecchia Olympic Village was part of a regional development plan that provided for financing the renovation and conversion of the entire 1930s complex.

Meanwhile, the Sestriere Olympic Village was built by a private company, which undertook the construction of the resort with the commitment to offer it free of charge to the organising committee during the Olympic event. Thus, Turin's transformations were emblematic of new post-industrial metropolises' regeneration and transformation processes. This strategy aimed to extend the benefits of the Olympic investment beyond the city, that is, to the entire region, thanks to the possibility of improving skiing facilities and developing the tourist season (Dansero, 2003). Thus, the dimension of the winter event in the fourth phase will turn into a metropolitan event that will inspire the reconstruction of the regional infrastructure.

Turin 2006

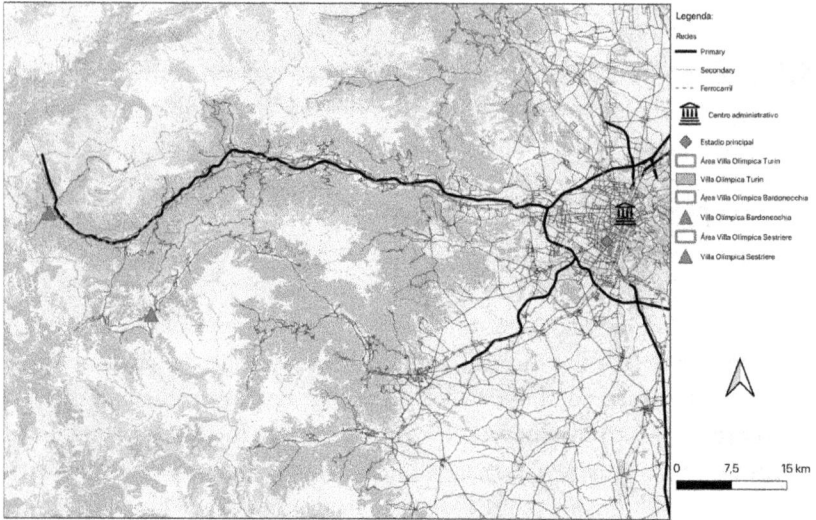

Figure 29. Spatial organisation in Turin 2006 (Source: Own implementation)

Subsequently, in Vancouver 2010, it advanced a new mixed financing model, introducing a new post-Olympic planning model that will promote long-term Olympic investment (VanWynsberghe, 2012). The Olympic housing planning was planned in an abandoned area that, thanks to private participation, could be completed, and new residential housing could be provided in the post-Olympic period. The new neighbourhood, consisting of thirty-seven buildings ranging in height from five to ten storeys, was reconfigured in the post-Olympic period and transformed into a central space in metropolitan Vancouver (COJO, 2010). However, the Vancouver Olympic Village will manifest other problems related to Olympic building speculation in the post-Olympic period (Scherer, 2011). During the construction phase, the goal of developing mixed-market housing was modified to provide only 10 per cent of the planned 30 per cent.

Meanwhile, in the post-Olympic period, rising rental prices increased evictions in the city (Essex, 2017). Subsequently, at the 2014 edition in Sochi, a spatial organisation was proposed that represented a milestone in expanding the new Olympic event to locations with a subtropical climate (Scott, 2015). The Russian event intended to develop a new spatial system by constructing new tourist sites and planning multiple Olympic villages connected with a railway system. After Sochi, the size of the event will continue to grow, becoming a strong stimulus for the transformation of the regional system even in places without snow. Undoubtedly, the 2014 event will cast new doubts on respecting and protecting the environment. However, like Turin, Sochi will propose a solution based on three Olympic villages: a main village near the ice facilities and ceremonial sites and two other Olympic villages in the mountain sites. Ninety-nine new buildings between two and seven storeys have been constructed (OCOG, 2014). The central Olympic Village will be converted into residences in the post-Olympic period, while the Olympic Villages in the mountains will be converted into hotels and resorts after the event, promoting tourism to the site. However, the post-Olympic edition has been criticised for substantial financial investments and interminable distances between Olympic venues. The Sochi edition raises new questions about post-Olympic development in sensitive mountainous or coastal areas. PyeongChang and Beijing 2022 will be other editions using the Olympic Village to promote sports tourism in mountainous locations. PyeongChang has provided a model of metropolitan clustering with a head office in the city and two Olympic Villages that will be used as residential accommodation in the post-Olympic period. However, residences in the mountain areas still need to be abandoned today.

On the other hand, Beijing 2022 will make the Chinese metropolis the first city in the world to host both the summer and winter editions. The Chinese edition will play a vital role in the metropolitan and regional dimension of the Winter Olympics. The spatial model is strongly inspired by the dimensions of Turin 2006, with the organisation of three Olympic Villages in a restricted territory. The central Olympic Village was built in an area bordering the Summer Olympic Village and, in the post-Olympic period, will be offered as a residence by public tender. The Zhangjiakou and Yanqing Olympic Villages in the mountain clusters were built to become tourist accommodations in the post-Olympic period.

Beijing 2022

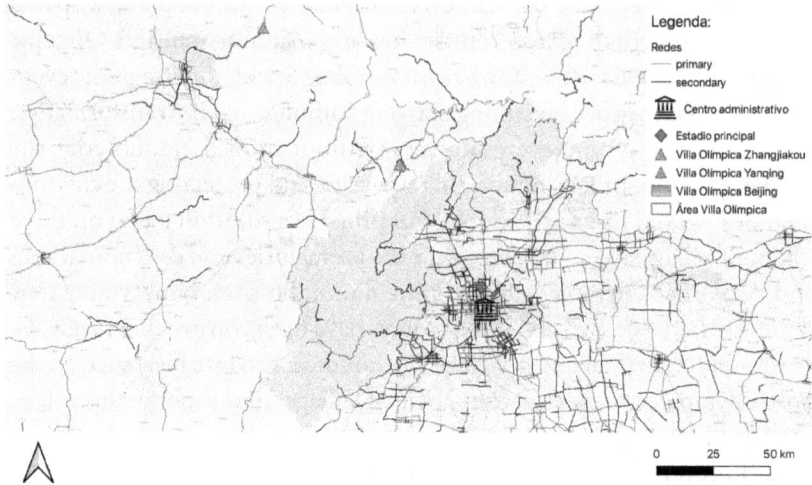

Figure 30. Spatial Organisation in Beijing 2022 (Source: Own implementation)

5.2.6 Phase 6: Tool for infrastructural development of tourism sites: Multiple cities, multiple regions (2026–Future)

In this last phase, the distance between the metropolis and the Olympic venues will reach an average of 115.63 km, promoting a new form of trans-regional Olympic development. The assignment of Milan-Cortina in 2026 and the possible assignment of Barcelona-Pyrenees in 2030 will mark a new planning scale. Beijing can only be considered the beginning of a new era for ephemeral bids, which will only use the metropolis as a promotional tool to secure the event. The metropolitan city will change to a place where tourism benefits and guest services will be exploited. Therefore, the Metropolitan Olympic Village has become a key element in the housing planning of the world's future metropolises, inscribed in the new urban dynamics of consumer societies. In addition, the Milan-Cortina 2026 edition will include three regions in north-eastern Italy by organising two principal cities and thirteen secondary venues. This dimension will lead to a new evolution of the winter event, turning it

Table 24. Stages of the Olympic Villages at the Winter Olympics (Source: Own implementation)

Phase	Years		
Phase I	1924-1948	Mountain locations Temporary accommodation	Prospects for the creation of an Olympic Village Existing sports facilities Use of hotels and resorts
Phase II	1952-1964	Cities with more than 100,000 inhabitants Permanent accommodation	Construction of the Olympic Village Different areas for the celebration of the event Developing a public policy for Olympic accommodation Growing interest in winter sports
Phase III	1968-1988	Regional expansion Residential accommodation	Encouragement for the creation of new sports facilities Development of the infrastructural system for the transfer of athletes. The foundations are laid for developing residential accommodation in the post-Olympic phase. New transformation model
Phase IV	1992-2002	Increase in Olympic space Tourism development tool	Increase in competitions and athletes Construction of multiple Olympic Villages New housing solutions (universities, demountable) Olympic space is organised in multiple locations. Respect for the environment

(*continued*)

Table 24. Continued

Phase V	2006-2022	Olympic Village in the city and Olympic Villages at competition venues Stimulus for the transformation of the regional system Metropolis	Main Olympic Village in the metropolitan city Mixed economy for the construction of residences in the mountain places The Olympic Village as a tool for the promotion of sports tourism in mountain areas Increased emphasis on environmental protection and sustainable development Legacy begins to enter into post-Olympic planning
Phase VI	2026 - Future	Multiple Olympic cities Multiple regions	Regional development Tool for the reorganisation of the economy of the Olympic area Creation of new mixed accommodation solutions Development of new infrastructure for the transport of Olympic athletes

into a new instrument for reorganising the national economy (the three Italian regions produce 1/5 of the GDP of the entire nation). In addition, developing new infrastructures will be a new challenge for the bidding cities. The candidature of Barcelona and the Pyrenees for the 2030 edition will make it possible to identify the new extraterritorial dimension that the Winter Olympics will be able to achieve. In conclusion, the new Olympic bids imply reconsidering the dimension of the event so as not to compromise the future evolution of our cities and territories.

5.2.1. *Spatial models of the Winter Olympic Villages*

As noted above, Winter Olympic Villages have historically been integrated into a territorial development policy and evolved as significant

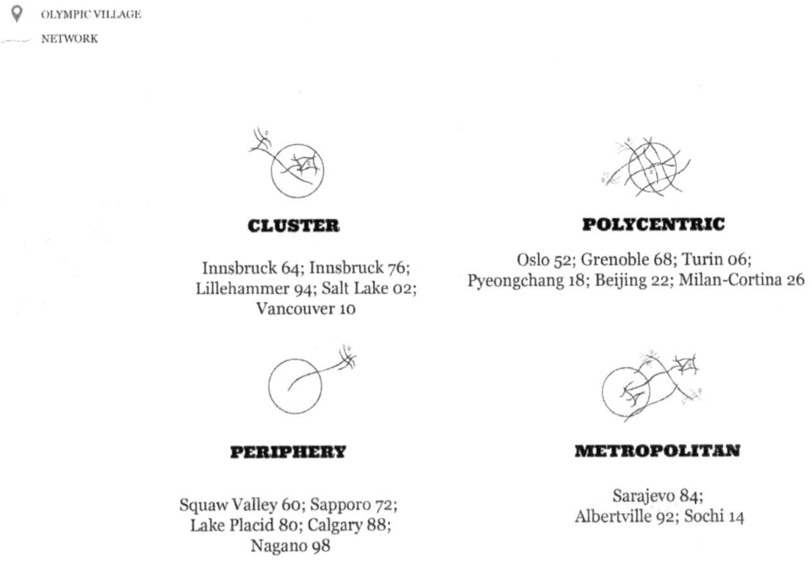

CLUSTER

Innsbruck 64; Innsbruck 76;
Lillehammer 94; Salt Lake 02;
Vancouver 10

POLYCENTRIC

Oslo 52; Grenoble 68; Turin 06;
Pyeongchang 18; Beijing 22; Milan-Cortina 26

PERIPHERY

Squaw Valley 60; Sapporo 72;
Lake Placid 80; Calgary 88;
Nagano 98

METROPOLITAN

Sarajevo 84;
Albertville 92; Sochi 14

Figure 31. Spatial models of the Olympic Village at the Winter Olympics
(Source: Own implementation)

elements of reconfiguring regional strategies. Generally, the model of
hosting athletes with competitions in mountainous areas is still indis-
pensable. However, the Olympic village model in mountainous areas continues to evolve through a new tourism development policy at Olympic
sites. Meanwhile, the polycentric model of the Candidate City has established itself as a critical element in the organisation of multiple Olympic
Villages. The decentralised model is one of the most widely used models
for the strategic development of the Olympic Village in the city.

However, recent experiences allow us to observe how the consideration
of the Olympic housing project in the central fabric of cities has evolved.
On the other hand, the peripheral cluster model is recognised as one of the
most widely used models for providing diverse housing in an expanding
territorial dimension. The analysis of Olympic Villages within regional
transformations can help to observe the evolution of the winter event dimension in the context of the regional expansion of host cities. Winter
Olympic cities have become global metropolises capable of proposing

new infrastructures for an ephemeral connection between the mountain space and the metropolitan city. Today, cities should reconsider the new dimension they have achieved and rethink solutions in a dynamic and open space that can be modified in a participatory manner. In the evaluation of the Olympic Village, no model or pattern can be identified, but a trend towards regional expansion of the world's largest metropolises can be observed (della Sala, 2022). However, climatic, thermal, and landscape constraints are noted for hosting winter events. Finally, the Sochi 2014 event proved that it is possible to organise a winter event in a subtropical climate with no historical value regarding sports facilities. Figure 31 shows the different spatial models that host cities have adopted.

5.3. Considerations on the urban strategies of the Olympic Village

The evolution and conception of Olympic town planning represented by the Olympic Village during the twentieth century allows us to observe the evolution of different urban, regional and metropolitan strategies that progressively increased the size and spatial characteristics of the first editions of the event. In addition, since the Second World War, the Olympic Village has become a key element in promoting housing permanently embedded in the host territory. However, the spatial dimensions of the two editions and the functions of the Olympic Village are different and continue to evolve about the housing needs of the host cities. Therefore, the Olympic Village can be defined as an element in its own right that not only functions as a place of temporary accommodation but is permanently inscribed in the territory in the different urban strategies of each host city. As we have seen in the previous sections, the Olympic Villages of the last two editions became private residences in the post-Olympic period. However, the construction of temporary housing or tourist accommodation is not the first choice of the organising committees.

Table 25. Examples of classification according to the character of each Olympic Village (Source: Own implementation)

SUMMER OLYMPIC VILLAGES		WINTER OLYMPIC VILLAGES	
PERMANENT	*TEMPORARY*	*PERMANENT*	*TEMPORARY*
1. Berlin 1936	1. Paris 1924	1. Oslo 1952	1. Lake Placid 1980 (prison)
2. Helsinki 1952	2. Los Angeles 1932	2. Squaw Valley 1960	2. Calgary 1988 (university)
3. Melbourne 1956	3. Tokyo 1964	3. Innsbruck 1964	3. Albertville 1992
4. Rome 1960	4. Los Angeles 1984 (university)	4. Grenoble 1968	4. Lillehammer 1994
5. México 1968	5. Atlanta 1996 (university)	5. Sapporo 1972	5. Salt Lake 2002 (university)
6. Monaco 1972		6. Innsbruck 1976	
7. Montreal 1976		7. Sarajevo 1984	
8. Moscow 1980		8. Nagano 1998	
9. Seoul 1988		9. Turin 2006	
10. Barcelona 1992		10. Vancouver 2010	
11. Sydney 2000		11. Sochi 2014	
12. Athens 2004		12. PyeongChang 2018	
13. Bejing 2008		13. Beijing 2022	
14. London 2012			
15. Rio 2016			
16. Tokyo 2020			

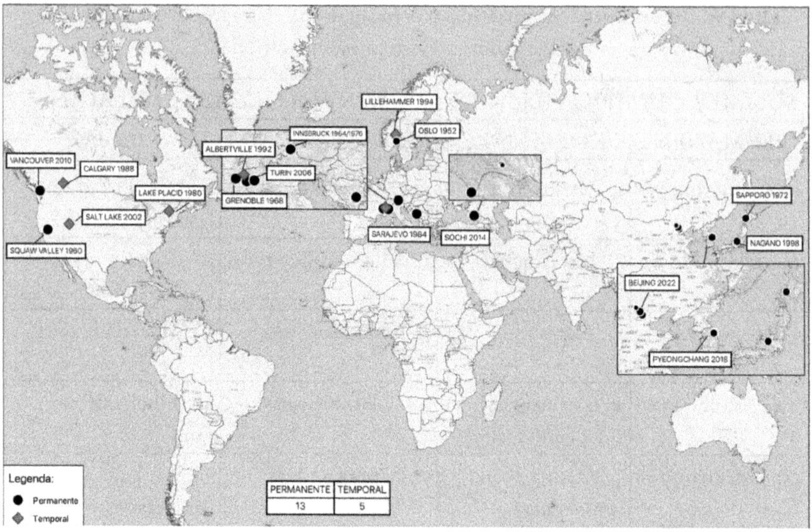

Figure 32. Classification of Winter Olympic Villages according to character
(Source: Own implementation)

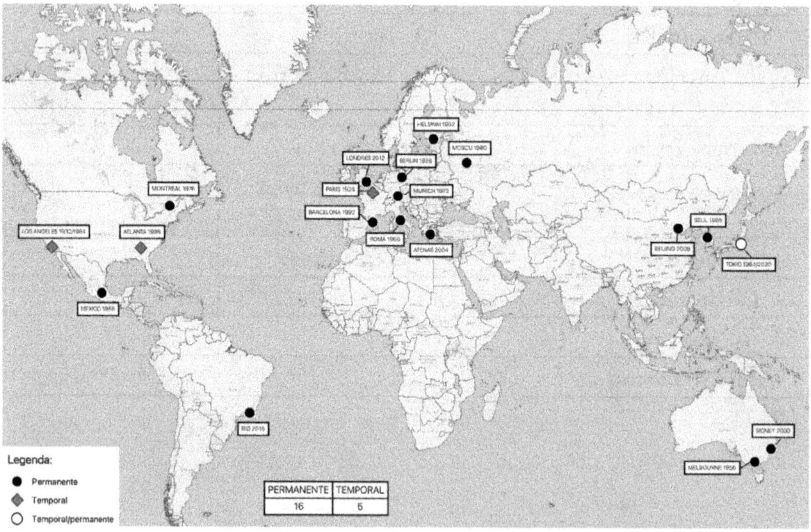

Figure 33. Classification of Summer Olympic Villages according to character
(Source: Own implementation)

Table 26. Examples of classification according to the ex-post evolution of the games
(Source: Own implementation)

SUMMER OLYMPIC VILLAGES		WINTER OLYMPIC VILLAGES	
REUSE	*ABANDONED*	REUSE	*ABANDONED*
1. Helsinki 1952	1. Berlin 1936	1. Oslo 1952	1. Sarajevo 1984
2. Melbourne 1956	2. Athens 2004	2. Squaw Valley 1960	2. Torino 2006 (inner city)
3. Rome 1960	3. Rio 2016	3. Innsbruck 1964	
4. Mexico 1968		4. Grenoble 1968	
5. Monaco 1972		5. Sapporo 1972	
6. Montreal 1976		6. Innsbruck 1976	
7. Moscow 1980		7. Lake Placid 1980 (prison)	
8. Los Angeles 1984		8. Calgary 1988 (University)	
9. Seoul 1988		9. Albertville 1992	
10. Barcelona 1992		10. Lillehammer 1994	
11. Atlanta 1996		11. Nagano 1998	
12. Sydney 2000		12. Salt Lake 2002 (University)	
13. Bejing 2008		13. Torino 2006	
14. London 2012		14. Vancouver 2010	
15. Tokyo 2020		15. Sochi 2014	

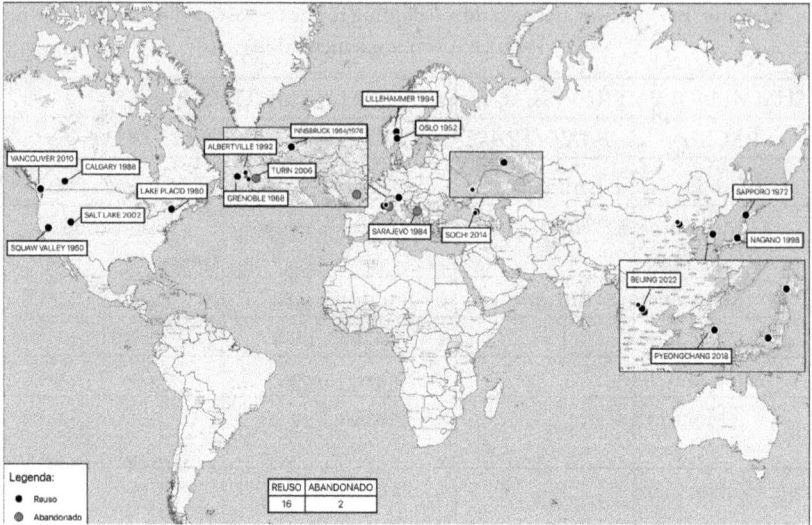

Figure 34. Classification of Winter Olympic Villages according to post-Olympic evolution (Source: Own implementation)

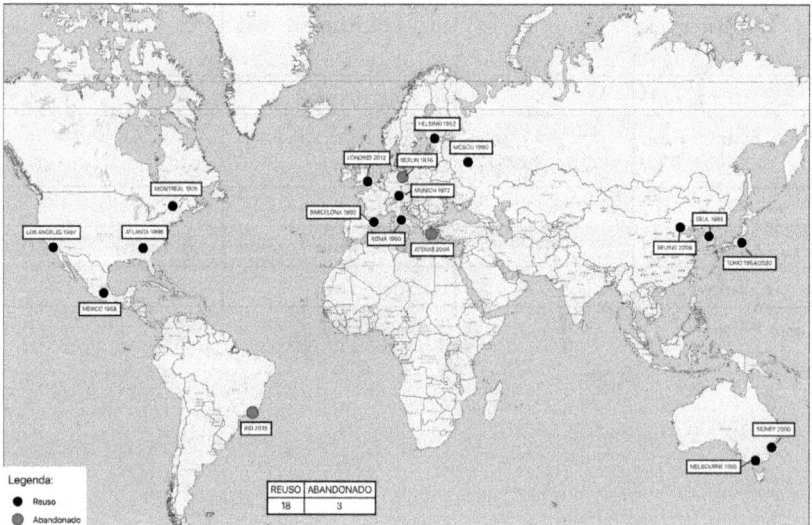

Figure 35. Ranking of Summer Olympic Villages according to post-Olympic evolution (Source: Own implementation)

Undoubtedly, after the experiences of Barcelona (1992), Turin (2006) and London (2012), the Olympic Village is an urban element that must be included in the strategic planning of the various programmes of reconstruction and redefinition of urban functions (della Sala, 2022). To conclude, through the adoption of cartographic representations, a classification of the Olympic Villages is advanced according to their character and post-Olympic evolution in terms of uses.

5.3.1. *The formal language of Olympic housing through the twentieth and twenty-first centuries*

The following section will observe an initial classification of the Olympic Villages built in the twentieth and twenty-first centuries. The different evolutionary stages of the Olympic Villages formal language will be listed through the analysis of building types and their formal language. As noted above, the first Olympic accommodations were conceived through single-family solutions or wooden huts reminiscent of the suburban dwellings observed in the nineteenth century. The philosophy of the garden city or suburban district was only introduced in the first decades of the twentieth century. Thus, spatial organisation in the first decades of the twentieth century began to explore a new scientific methodology in the urban planning of our cities. Subsequently, the concept of multi-family dwellings built according to the rationalist theories of CIAM was introduced. Thus, the Olympic project's approach to the city centre involved reconsidering spaces and reconfiguring abandoned ones. For example, the layout of the Olympic lodgings during the 1972 Munich edition is an excellent example of the following type of development.

On the other hand, the Olympic Villages of Mexico 1968, Moscow 1980 and Rio 2016 are characterised by the construction of large housing complexes based on the implementation of peripheral projects organised in housing blocks. Compared to the other groups, the Olympic Villages of Moscow in 1980, Seoul in 1988 and Barcelona in 1992 were included in a city renewal plan. Meanwhile, the Olympic Villages of Los Angeles 1984, Calgary 1988, Atlanta 1996, and Salt Lake 2002 will be ephemeral projects for the Olympic event. However, they will be catalysts for other processes within the host universities.

5.3.2. Satellite city and garden city

The Olympic Villages in Los Angeles in 1932 and Berlin in 1936 clearly show the implementation of the garden city philosophy for the urban and territorial organisation of the metropolis' peripheral space. Furthermore, the Berlin Olympic Village offers elements of modernity by applying the principles of the Bauhaus school. On the other hand, Los Angeles presents a housing unit that respects the significant prefabrication concept of the first decades of the twentieth century.

Therefore, the idea of the Berlin Olympic Village allowed for reflection on the rationalisation of interior and exterior spaces through the conception of efficient and adequate housing.

5.3.3 Rationalism and functionalism

After the post-World War II experiences, characterised by the ephemeral and the rapidity of realisation, the 1960s will represent a fundamental moment for applying the functionalist ideas of Le Corbusier and CIAM. The separation of functions on the territory and the importance of housing, green spaces, commerce, health, and sports facilities will be the elements that will determine the evolution of the building types of the Olympic Villages in Rome, Mexico, and Moscow. The Olympic Villages of the 1960s will be planned through modern town planning proposals, characterised by new elements such as leisure and public space (Modrey, 2008). In the case of Rome in 1960, the Olympic Village introduced the functionalist aspects of the greatest Italian architects, adding the pylons and defining the height and regularity of the façades. In addition, the rationalist theory allowed for considering other fundamental parameters such as air quality, light conditions, and the organisation of interior space. Therefore, the Olympic Village in Rome can be defined as a small *Villa*

Radieuse that introduced fundamental architectural and urban planning elements used in the planning (della Sala, 2022). However, Mexico's 1968 Olympic Villages represented a new modernity strongly influenced by European proposals on housing and the transformation of space in central areas previously occupied by factories. Therefore, the Mexico project reflects the ambition to develop new self-sufficient neighbourhoods in peripheral areas using standardised construction techniques. Finally, it should be noted that the proposals for Moscow in 1980 and Rio in 2016 reflect the modernist typology observed in Mexico in 1968.

5.3.4. Centrality and radicalism

Given the Olympic Villages of the 1970s, several applications in contemporary architecture can be observed. The new visions of urban design and the reorganisation of the centralities of major European cities will introduce a new vision. From this moment on, the reorganisation of urban space will include leisure areas, communal services, sports facilities and residential housing. Thus, Olympic planning begins expanding and manifesting through new building models. The concept of the leisure city and the redevelopment of central spaces can be clearly observed in the Olympic Villages in Munich and Montreal. Munich's 1972 accommodation was integrated into a vast urban park. Montreal 1976 prepared a new Brutalist quarter that included all functions in four large pyramid complexes. Thus, both Olympic Villages were integrated into urban tasks while retaining their autonomous characteristics.

5.3.5. Postmodernity and the ephemeral place

As noted above, the Olympic Villages of the 1980s and 1990s formed the basis for new interventions and the implementation of different forms of expression. The accommodations provided for Seoul, Barcelona and Moscow have many elements in common regarding urban renewal and strategic planning (della Sala, 2022). The realisation of new Olympic

Villages in an era of postmodernity allows local elements to merge with global ones to be massively incorporated into the city (Dickson, 2005). Moreover, in the postmodern era, the Olympic Village is configured as a central element in the main urban structure of host cities and as a catalyst for other proposals. Therefore, the architecture of the Olympic Village becomes a new proposal for the transformation of place by time, new forms and architectural technologies. In this postmodern phase, Olympic Villages will be strategically planned in the central fabrics of the host metropolises. Only in Athens 2004 and Rio 2016 will we observe two formal expressions reminiscent of past experiences that are not the best option for urban regeneration and post-Olympic use.

The 1984 Los Angeles and 1996 Atlanta Olympic Villages represent a recycling solution for the infrastructure available for Olympic operations. As noted earlier, Olympic Villages have evolved throughout history using different models and construction types in other areas of the host cities. However, the structural changes in each city allowed for certain practices to be implemented to understand the leisure and recreation needs of the population after World War II.

In conclusion, the different models of Olympic Villages observed can help us reflect on the development of our cities and the evolution of places of consumption by athletes and citizens. The garden city or the suburban city are not solutions in line with Olympic times and the ideas of postmodernity. The abandonment of Athens and Rio helps us appreciate the importance of post-Olympic planning in order not to leave abandoned spaces in the era in which we live, as the right to housing is fundamental. However, the evolution of the event and the spectacularity of sport have provoked a rethinking of the main spaces and the construction of new places of consumption in the historical areas of cities. The experiences of Los Angeles in 2028 and Paris in 2024 represent the city's alter ego. A small ephemeral, demountable and temporary dimension that, over the years, has turned into an element with its own urban identity, capable of catalysing new constructions and new housing proposals. Thus, in the future, we will see a new struggle between citizens and planners for consideration and community participation in planning.

CHAPTER 6

Case Study

As noted above, the spatial dimension of the Olympic event has undergone a great transformation over time. However, the Turin 2006 edition will mark a profound change that will imply a rethinking of the Olympic project by the host cities. Therefore, the following section will observe five emblematic case studies that have profoundly transformed the host territories.

The analysis of the case studies of Turin 2006, London 2012, Sochi 2014, Rio de Janeiro 2016 and Beijing 2022 will allow us to observe the Olympic Villages through the models proposed by della Sala in his doctoral study (see della Sala, 2022).

The following models were designed using a comparative analysis of certain statistical elements that allowed each Olympic Village to be observed through standard parameters.

For the first time in Olympic history, the Olympic Villages can be observed with objective criteria.

However, three winter and two summer editions were chosen because of their importance in the evolution of the spatial dimension of the Olympic event.

As noted in Chapter 4, Turin 2006 catalysed a transformation of regional space, allowing future candidate cities to expand their municipal boundaries.

Therefore, the design of three Olympic villages during the Italian edition was also proposed for the Sochi and Beijing editions. The London and Rio de Janeiro events were selected to analyse the summer editions. London because of its importance in the revaluation of disused spaces within central areas. Because of its peripheral spatial organisation, Rio de Janeiro introduced new elements related to the mobility and transport of athletes.

Therefore, we could compare two events with different dimensions through a series of standard parameters by analysing both summer editions.

Finally, after observing the five case studies, we will proceed with conclusions regarding the spatial dimension and territorial implications of these events within our territories.

6.1. The Olympic Villages of Turin 2006

As noted above, the Turin 2006 Olympic event was organised through a new spatial model that included different areas to define the competition venues.

As such, the Organising Committee proposed an accommodation solution by providing three Olympic Villas. The Villas of Torino, Bardonecchia and Sestriere. In the following sections, we will examine the history and evolution of the three accommodation complexes post-Olympic period.

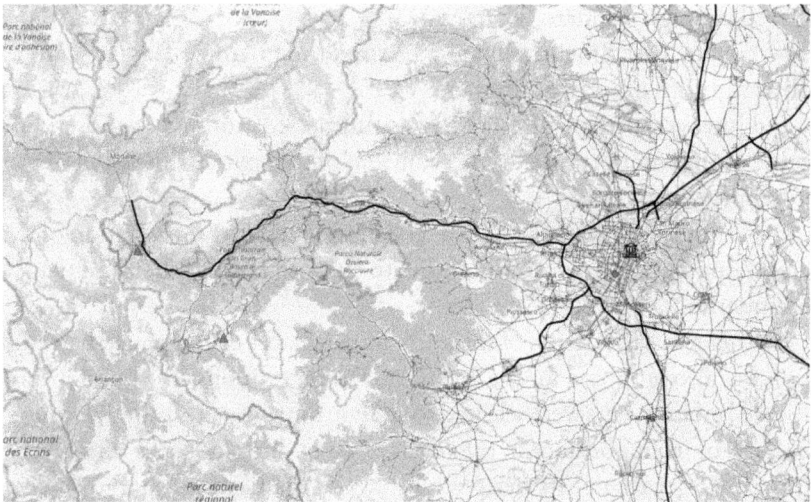

Figure 36. Location of the Olympic Villages in Turin 2006 (Source: Own implementation)

6.1.1 *The Turin 2006 Olympic Village*

The Turin Olympic Village is considered the central Village, located in the urban context of the metropolitan city. The location of the Villa was chosen for its proximity to the main sports facilities and for its integration into the Lingotto area, which at that time was defined in the 1995 master plan as an area subject to post-industrial revaluation. The work was part of the metropolitan redesign, integrating into the urban fabric and bringing new values and identity into a historical context where the old General Markets, designed by Umberto Cuzzi in the 1930s. Therefore, the Olympic Village was conceived as a mixed residential complex in a strategic area for connecting the city's southern area. In the 1998 bid dossier, the Turin Olympic Village was defined as a pilot project destined to become an example of sustainable architecture. Furthermore, the complex was one of Europe's first SEA (Strategic et al.) application cases. After the public tender in 2002, the group of builders represented by the architect Camerana (architect of the *Agnelli* family) proposed a solution by designing a new urban envelope along the railway line.

The Olympic Village area consists of recovering the former wholesale fruit and vegetable markets (MOI) destined to house services and with a not well-defined hypothesis of reconversion. However, in the post-Olympic period, the city council did not propose any solution for reusing the service area.

For its historical-architectural importance, the MOI was considered the central element of the connection between the two areas separated by the railway line. The intervention strategies of the MOI were applied through a recovery that has maintained the skeleton of the exMOI, restoring the structures and obtaining new volumes (De Pieri, 2008).

The Olympic *walkway* connecting the Lingotto complex to the Olympic Village fits the MOI's centre of gravity. Meanwhile, the 657 residential units were distributed in 39 buildings conceived to become private residences in the post-Olympic period. The buildings feature commercial premises on the ground floors, common areas and services. In addition, after the Olympic event, ARPA[1] will set up its offices in lot 4. Subsequently,

---

1   Agenzia Regionale per la Protezione Ambientale del *Piemonte* (Regional Agency for the Environmental Protection of *Piedmont*).

Figure 37. Turin, Olympic Village, current state (Source: Personal archive)

the sports federations and the post-Olympic foundation will be installed in other blocks.

However, the conversion to tertiary uses, which was not foreseen in the initial project, required new investments to adapt to the new function. The sockets, the ventilation system, the layout of the interiors and the sanitary facilities had to be dismantled to adapt the spaces to office use. Moreover, in the post-Olympic period, the Village was occupied by refugees due to the uncertainty of the real estate market and the high cost of flats. Only part of the residences have been sold to private individuals and used as university halls of residence. Even though in 2015 the Judicial Authority ordered the seizure of the occupied buildings, today, the buildings remain abandoned, and their fate is uncertain.

The lack of specific functions and compatible uses for the spaces has led to the abandonment of the area, with numerous degradation phenomena related to inefficient execution and cheap materials.

However, the planning of the post-Olympic use was not correctly defined, as the objective of using the area as a part of the urban transformation strategy for the southern area was confused with the performance strategy itself.

Another critical aspect is the need for more maintenance and an apparent deficiency in the execution of the works, such as windows that cannot be opened, thermal heating not connected or load-bearing walls and stairs inside the dwellings. These are just some of the execution defects.

In the context of the rapid urban transformations in connection with the Olympics, the Turin Olympic Village is the most important work from a symbolic point of view.

> A new piece of city that replaces another, in front of the Lingotto. (Bianchetti, 2005)

6.1.2 The Olympic Village of Bardonecchia

The accommodation complex in Bardonecchia was included in a project to redevelop and reuse a 1930s colony. The complex was in an area with different buildings, including dormitories, toilets and shared spaces. The rooms were conceived as divided dormitories that included washbasins in each room. Like many of the works built by fascism in Italy, the complex had its central tower that made the colony visible from a distance. In addition, the colony has a refectory, a reading room and a recreation room. The "IX Maggio" mountain colony[2] was modified several times to transform it into an Olympic Village and to meet the standards set by

---

2   The colonies were sponsored during Mussolini's fascist government; in Italy, the
     colonies are considered cultural heritage.

Figure 38. Turin, Olympic Village, current state (Source: Personal archive)

the IOC. The ground floor of the building was completely altered, and another floor of rooms was proposed.

The double, triple and quadruple rooms are distributed on different levels and are equipped with every comfort. The Olympic Village complex includes large areas dedicated to sports and recreational activities, with a swimming pool with sauna, Turkish bath, whirlpool bath and gymnasium. The complex also includes two restaurants, a bar, large recreation rooms, a theatre, a discotheque and a lounge bar. An outdoor car park, a garage and a heated ski boot storage area complete the area. At the end of the Olympic event, the structure was assigned to the municipality of Bardonecchia

Figure 39. Bardonecchia, Olympic Village, present state (Source: Personal archive)

to increase the number of accommodations available in a tourist resort. Therefore, including the Olympic Village in a mountain village, proposed from the beginning as a tourist accommodation structure, has facilitated its post-Olympic use without developing a negative legacy for the territory.

Today, the Olympic Village is still fully operational through a public agreement between the municipality and a private hotel cooperative. The Olympic spirit of the Village is still alive in the large spaces of the complex.

6.1.3 The Olympic Village of Sestriere

Sestriere is the highest mountain resort in Italy, situated at 2035 m.a.s.l. and one of the most emblematic places for mountain competitions. Throughout its history, the town has hosted many editions of the Ski World Cup, standing out as one of the most important places in Europe.

The village of Sestriere has a recent history; its evolution is mainly due to the *Agnelli* family, which, through the construction of two towers, has been the propellant of a touristic construction process that will change the landscape of the place. Other towers were conceived as a single endless chamber through a helical development. The long ramp is one of the particularities of the construction of the Towers, where their singularity and uniqueness to the world stand out. Subsequently, Sestriere became a vital winter resort, attracting a high-level international clientele thanks to the modernity and style of its facilities. Sestriere is a historic venue for modern skiing, from *Alberto Tomba*'s double in 1987 to the first World Cup night slalom in 1994, the World Championships in 1997, the Alpine World Cup Finals and the 2006 Olympic Games. The resort is a successful fusion of modernity and tradition. The buildings are arranged along the contours of the terrain. The glass galleries and interiors connect the buildings without leaving the entire complex. The central body of the Olympic Village has a large service area and three floors of rooms and apartments.

The Olympic Village is a hybrid structure consisting of different types of rooms and flats. The complex also has a gymnasium and a health centre. From the beginning, the structure was financed by a private company, which, in the post-Olympic period, was added to the company. The collaboration and cooperation between the municipality of Sestriere and the company made it possible to provide a new accommodation complex, which, in the post-Olympic period, would be added to the availability of the mountain resort. In this way, the Sestriere Olympic Village can be considered a case study for the organisation and planning of a tourist complex in mountain resorts.

6.1.4 *Turin 2006*

The 2006 edition of the Turin Olympics is recognised as the first to include three different Olympic Villages in a territorial space that had never been contemplated before. The construction of an Olympic Village in the city and two other Olympic Villages in the mountain communities will be the catalyst for transformations in the dimensions and organisation

Figure 40. Sestriere, Olympic Village, current state (Source: Personal archive)

of the Winter Olympic event. The project envisaged constructing and re-using existing structures to accommodate individuals, students, and tourists. The Olympic Village in the metropolitan city of Turin was in a historic area that was converted into accommodation to become a new residential neighbourhood in the post-Olympic period. The construction of the Olympic Village was environmentally friendly through the use of reused materials, solar panels and independent heating[3] The Turin Village was close to most of the sports facilities implemented in the city, and the neighbourhood had all services and means of transport available[4].

Meanwhile, the other two Olympic Villas, in Sestriere and Bardonecchia, were planned to be converted into hotel accommodation

3 XX *Olympic Winter Games: Final Report*, Organising Committee for the XX Olympic Games, Winter Games, 2007, vol. 1, pp. 105–110.
4 "Village olympique de Torino", sites de compétition – Torino, website of Turin 2006.

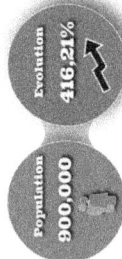

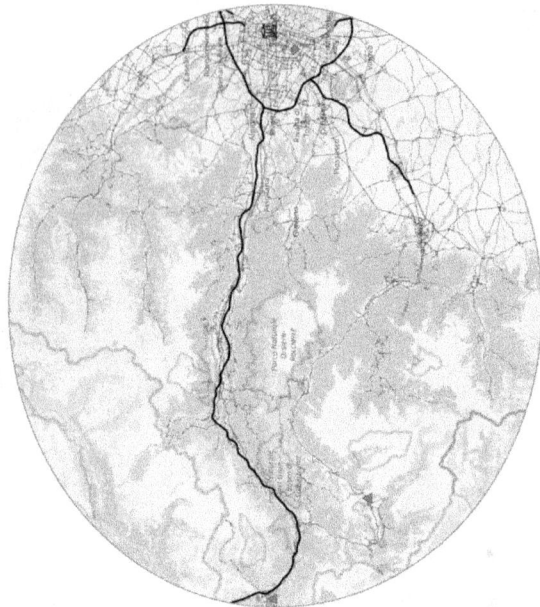

Figure 41. Model of the Olympic Village in Turin 2006 (Source: Own implementation)

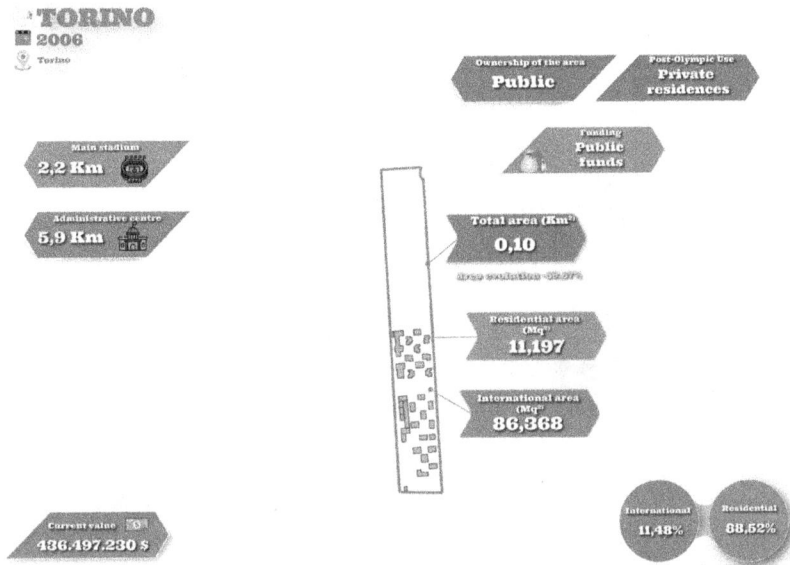

Figure 42. Model of the Olympic Village in Turin 2006 (Source: Own
implementation)

in the post-Olympic period. The Sestriere Villa was an extension of an existing structure and was sold to private individuals[5].

Meanwhile, the Olympic Village in Bardonecchia[6] was an abandoned hotel complex from the 1930s that was converted and completely renovated. In the post-event period, the aim was to convert the Turin Olympic Village into a mixed accommodation site, which to this day remains derelict. Meanwhile, the Olympic Villages of Sestriere and Bardonecchia planned to be hotels, have enabled the towns to provide new hotel accommodations for winter tourism. The Turin edition promoted a new type of development and exploitation of the Winter Olympics.

5   "Le Village olympique de Sestrières", sites de compétition – Sestrière, website of Turin 2006.
6   "Le Village olympique de Bardonecchia", sites de compétition – Bardonecchia, website of Turin 2006.

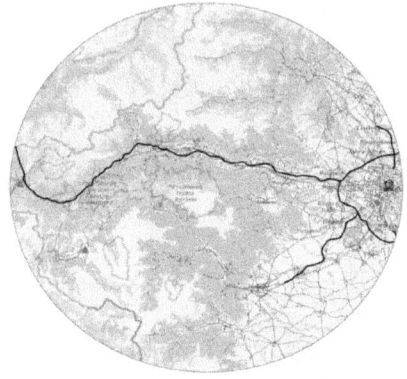

TORINO
2006
Bardonecchia

Number of buildings	Height of buildings	Typology of buildings
1	**4-5** Storey	**Hotel**

51,420 Occupied Surface m²
8,75% % Density m² (Resid)
12842 Urban Density (ab/area km²)

Population **900,000** Evolution **416,21%**

Surface **130,17 Km²** Capacity N.of athletes **2,508** Evolution **4,54%**

Figure 43. Model of the Olympic Village in Bardonecchia 2006 (Source: Own implementation)

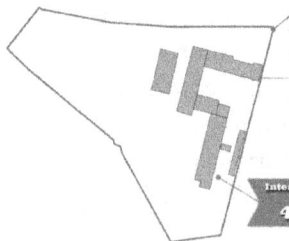

TORINO
2006
Bardonecchia

Ownership of the area **Public**   Post-Olympic Use **Hotel**

Main stadium **94,1 Km**

Administrative centre **89,4 Km**

Funding **Public funds**

Total area (Km²) **0,05**
Area evolution -99,97%

Residential area (Mq²) **7,238**

International area (Mq²) **44,182**

Current value **436.497.230 $**

International **14,08%**   Residential **85,92%**

Figure 44. Model of the Olympic Village in Bardonecchia 2006 (Source: Own implementation)

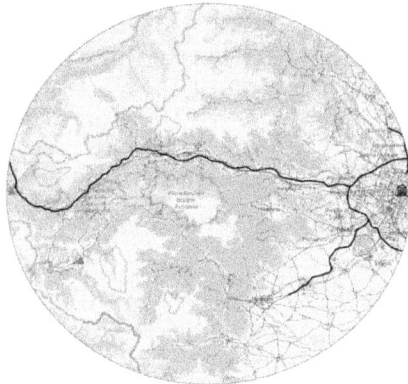

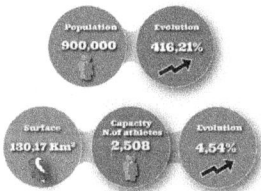

Figure 45. Model of the Olympic Village in Sestriere 2006 (Source: Own implementation)

Figure 46. Model of the Olympic Village in Sestriere 2006 (Source: Own implementation)

6.2. The Olympic Villages of London 2012

The London 2012 Olympic event was developed by providing an Olympic Village within the East Village area. The entire area was converted for use as a residential neighbourhood in the post-Olympic period. Thus, the London 2012 project is based on the urban regeneration of large abandoned spaces and derelict industrial buildings in the centre of Stratford.

6.2.1 The London 2012 Olympic Village

"Our Village is compact, but it feels spacious because of how the facilities have been laid out, the large central grass area and the Olympic Park

Figure 47. Location of the Olympic Villages in London 2012 (Source: Own implementation)

alongside – you can walk to all the venues in the Park. Every part of the Village is easy to reach – most facilities are no more than 400 m from your accommodation" (Sainsbury, 2012).

The London Olympic Village was designed as part of an extensive urban regeneration programme based on reusing buildings and creating a new residential quarter for Stratford. In addition, the whole area was included in an SCDC regeneration plan with the consequent creation of the London-Paris high-speed link. The execution of the project was carried out through a tendering procedure entrusted to Lend Lease[7], which included the financing and construction of the Olympic Village and part of the media centre.

The Olympic Village comprises 2,818 townhouses and flats and can accommodate 15,000 residents in the peak period. Each flat has been equipped with comfortable spaces and state-of-the-art communication facilities. In addition, wireless Internet access has been provided throughout the area. Thus, the Olympic Village fits within the large Olympic Park and is in a prime location for visiting the city of London. The design of the Village echoes the London tradition of building dwellings around communal squares and courtyards in a harmonious setting surrounded by water features. In addition, Olympic accommodation includes a wide range of commercial and recreational facilities adjacent to the Stratford City complex. The new high-speed Javelin shuttle service serves the entire area, connecting the Village to central London from Stratford International Station in just seven minutes.

The project was designed to meet Life Time Home standards to support Paralympic residents and future residents with disabilities or other physical difficulties. The entire area includes offices, shops, schools and a health centre. Lend Lease hired a team of architects, Fletcher Priest, structural engineers, Arup, and city planning and landscaping companies, West eight and Vogt Landscape, for the construction. The project involved constructing fourteen residential lots, five to seven blocks, built around large squares and inner courtyards. The squares and courtyards were equipped

---

7 Lend Lease is an Australian public multinational company responsible for constructing and selling real estate.

Figure 48. London, Olympic Village, current state (Source: CC)

Figure 49. London, Olympic Village, current state (Source: CC)

with water features and ample green spaces. Each of the sixty-nine blocks is between eight and twelve storeys high. There are primarily three-storey terraced houses at street level with street-level entrance doors to create an active façade. The following blocks are joined by shops, offices and services ranging from one to three storeys. On the third floor of the complex are raised communal gardens. In addition, each flat has its balcony to promote the relationship between indoors and outdoors.

The entire area of the Olympic Park was proposed for compulsory purchase (CPO) by the London Development Agency. However, at the end of 2005, a controversy broke out between the Mayor of London, Ken Livingstone and the Newham Council/Westfield Group regarding the legality of the acquisition instrument. Therefore, in light of the 2008 financial crisis, Lend Lease found raising funds on the markets to construct the Olympic Village challenging.

At this point, the government, through the Olympic Delivery Authority (ODA), agreed to underwrite more than the required sum and impose a scaling down of the project by 25 per cent. One of the conditions stipulated by the ODA was the postponement of the construction of three lots (5–6–8) after the end of the event. However, the following failed to secure the beds for the Games period. Therefore, a design plan for the event introduced time walls to create more beds. In this way, a two-bed flat with only one bedroom was transformed into a four-bed flat. The following intervention allowed the division of the living, dining and kitchen areas into an open plan. After completion, everything was readjusted to the original concept to facilitate the accommodation sale to private individuals.

Until then, the Olympic site was a mix of former industrial buildings and contaminated land. There were two campsites for Irish Travellers in the Olympic Village area, one in Clays Lane, Newham, and another in Waterden Crescent, Hackney. In 2006, the Travellers had unsuccessfully tried to appeal to the High Court in May 2007.

In addition, towards the end of the demolition and clearance of the entire area, a fire broke out in an old industrial warehouse in Waterden Road, Hackney Wick, on 12 November 2007, where it took seventy-five fire brigades to extinguish the flames another 15 m.

Figure 50. London, Olympic Village, current state (Source: CC)

As noted above, after the Games were over, the housing was adapted to create an East Village housing estate. The complex, consisting of 2,818 dwellings, included 1,79 homes and affordable housing for sale and rent by arrangement. In 2009, the ODA sold the affordable homes to Triathlon Homes for £268 million. Meanwhile, in 2008, the ODA launched a tender to sell the remaining 1,439 private homes and six adjacent development plots with building potential for a further 2,000 new homes and the long-term management of the entire East Village area.

In August 2011, the ODA announced a deal with the Delancey/Qatari Diar fund that paid £557 million for the purchase of the entire East Village site, foreseeing an estimated loss of £275 million for the ODA and thus for the Culture Secretary Jeremy Hunt commented at the time that the ODA did not expect to recoup the construction costs because the whole area needed infrastructure, roads and parks, so the public contribution was crucial for the construction of the new district.

After the Olympic event, the hotel-style designed rooms were con-verted to include a lounge and kitchen. The 1,439 private houses were rented instead of sold, and the property remained in Delancey/Qatari Diar and was managed by Get Living London. The following choice fostered the creation of the UK's first private residential fund of over 1,000 homes owned and managed directly by the fund as a speculative investment.

The promoters also added the creation of the Chobham Academy, a new educational campus with 1,800 places for students aged 13–19. The entire school building was used as the main base for the organisation and management of the teams during the Olympic event.

In conclusion, the planning of the post-Olympic use was adequately defined by including private parties and creating a real estate fund. However, as noted, the Olympic Village was part of an area included in the urban transformation strategy of the Eastern area.

6.2.2 London 2012

The London Olympic Village project was part of a massive transformation of an industrial area in the eastern part of the metropolis. The Olympic Park and Olympic Village comprised an area of more than 10 hectares that could accommodate more than 18,000 athletes. The Olympic Village was built to be converted into a mixed housing area, conceived with the utmost respect for the environment. However, the London project was the first example of applying principles that respected the concept of heritage and Olympic heritage. The proximity to the Olympic Park made it possible to reduce transport services between the training facilities and the Olympic Village. The area was planned to accommodate shopping centres, train stations, underground, cinemas, nightclubs, theatres, offices, consultancies, restaurants, bars, gymnasium and other services. The London Olympic Village reached unprecedented dimensions. Athletes in the sailing and equestrian competitions were accommodated in hotels. The project was planned through a public-private partnership to exploit the entire area (East Village) and develop it into a new neighbourhood, which would be very close to the city's administrative centre. The London Olympic Village will be the first example of mixed accommodation and services built through the summer edition.

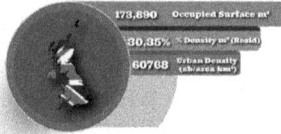

Figure 51. Model of the Olympic Village in London 2012 (Source: Own implementation)

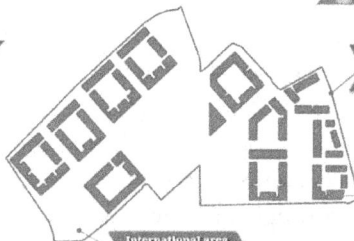

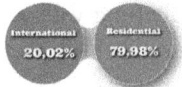

Figure 52. Model of the Olympic Village in London 2012 (Source: Own implementation)

6.3. The Olympic Villages of Sochi 2014

The Sochi 2014 Olympic event was developed by constructing a huge Olympic Park. The urban settlement of Sirius in the Imeretinsky Valley, on the Black Sea coast, included the central Olympic Village, the Olympic Stadium and other main competition facilities. In addition, the Park houses training facilities, a media centre and other services. It was designed so that athletes could walk to all the facilities. However, two more Olympic villages were set up at the competition venues arranged in the mountain cluster to meet the event's needs. The first village is in Rosa Khutor, and the second is in Sloboda.

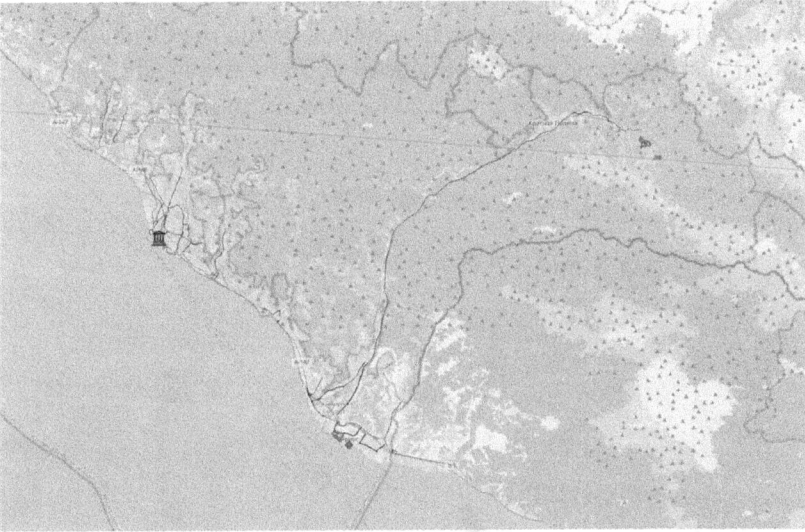

Figure 53. Location of the Olympic Villages in Sochi 2014 (Source: Own implementation)

6.3.1 The Sochi 2014 Olympic Village

The central Olympic Village of the 2014 Sochi Winter Games was harmoniously embedded within the Olympic Park area. The construction of the central accommodation was entrusted to the company Basic Element[8].

The main village consists of 47 buildings designed to accommodate 3,000 people. Athletes, officials and technical staff were accommodated during the event. The accommodations were designed to be resold as private flats in the post-Olympic period.

Thanks to positive relations with investors, the organisers could complete the accommodations on time and comply with security guidelines.

The Sochi Olympic Village is a mega-project that has attracted attention beyond the mere sporting event. The design of the entire area included the construction of a highway, railways, access to the sea, power plants, recreational facilities, etc. In addition, within the Olympic area is a 3-star aparthotel and a 4-star hotel with 3,600 rooms, an administrative building, sports complexes and social facilities on an area of 120,000 m².

Figure 54. Sochi, Olympic Village, current state (Source: CC)

8   Basic Element Ltd is one of Russia's largest diversified industrial groups. The company was founded in 1997 and is owned by Oleg Deripaska.

6.3.2 *The Rosa Khutor 2014 Olympic Village*

The Rosa Khutor Alpine Resort is an alpine ski resort in Krasnodar Krai, located on the Aibga ridge of the West Caucasus, along the Roza Khutor Plateau, near Krasnaya Polyana.

In addition to hosting Alpine skiing competitions, the following ski resort was also the site of the Olympic Village in the mountain cluster. In addition, the Alpine ski resort was the site of construction ten hotel projects, creating 1,600 rooms to exploit winter tourism in the post-Olympic period. The entire resort could accommodate 2,900 people and was 50 km east of the Black Sea in Sochi.

6.3.3 *The Sloboda Endurance Village 2014 Olympic Village*

The accommodation was designed following a mixed philosophy in Sloboda's second mountain village, the Olympic Endurance Village.

Figure 55. Rosa Khutor, Mountain Village, current state (Source: CC)

Therefore, the facilities were designed to accommodate 1,100 athletes by providing hotels, flats and chalets. Through the provision of mixed accommodation, the Organising Committee wanted to encourage the emergence of a new mountain cluster that could bring together winter sports enthusiasts in the post-Olympic period.

In addition, the proximity to the ski resort and the cross-country skiing centre allowed a quick connection to the other competition venues during the Olympic event.

The entire village is set in a stunningly beautiful environment, and the architecture is reminiscent of Russian folk tales. In addition, each village has an Olympic mascot reminiscent of a specific character from Russian fairy tales, such as Baba Yaga, the Firebird, Ilya Muromets, Sadko, Mad Ivan and the Grey Wolf.

Figure 56. Sloboda, Endurance Village, current state (Source: CC)

6.3.4 Sochi 2014

The organisation of the Sochi Olympics was embedded in a very complex spatial planning where the three Olympic Villages were the three main centres of this spatial model. Each Olympic Village had its own identity and its utmost post-Olympic function. The Sochi Village was in a new area close to the venues and sports halls. In addition, the Village was included in an area with all kinds of services. The other Olympic Villages were arranged in Rosa Khutor and Sloboda; they had their own identities and the same level of service as the Sochi Olympic Village[9]. The accommodation in the mountain villages in the post-Olympic period was converted into hotels and flats. Meanwhile, the Sochi Olympic Village was reused as a residence. The Sochi edition is considered inspirational for

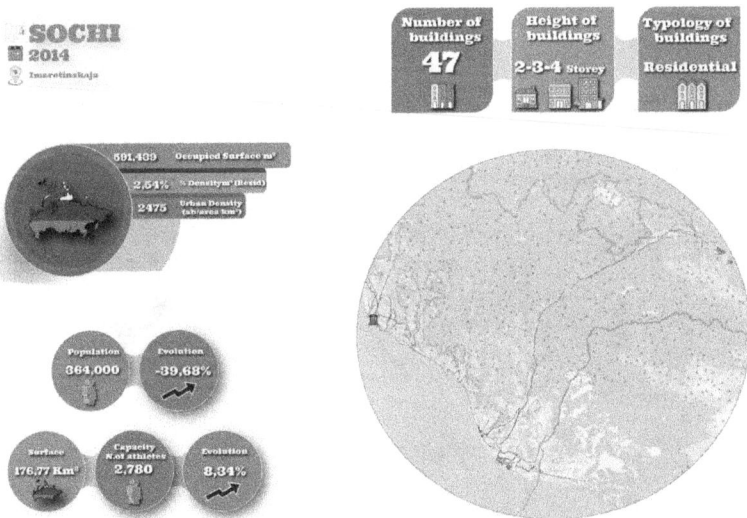

Figure 57. Model of the Olympic Village in Sochi 2014 (Source: Own implementation)

9   *Rapport Officiel: Sochi 2014 Games Olympiques d'hiver*, Sochi Organising Committee for the 2014 Olympic and Paralympic Winter Games, 2015, vol. 3, pp. 36–38.

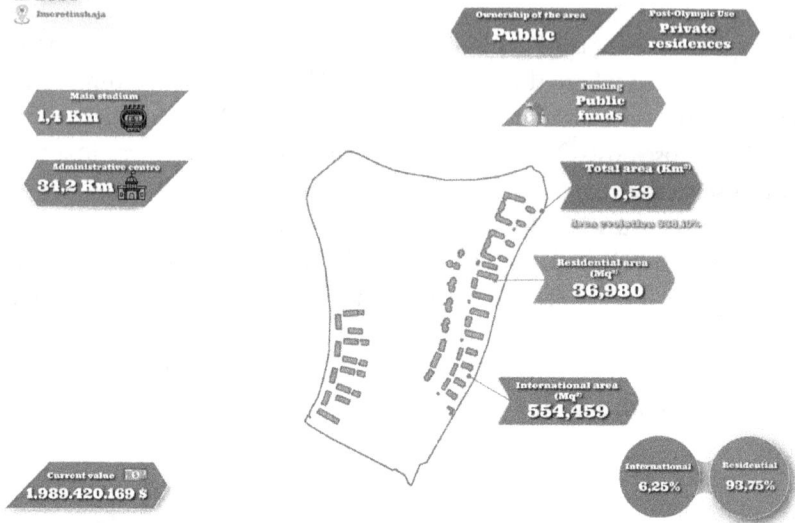

Figure 58. Model of the Olympic Village in Sochi 2014 (Source: Own implementation)

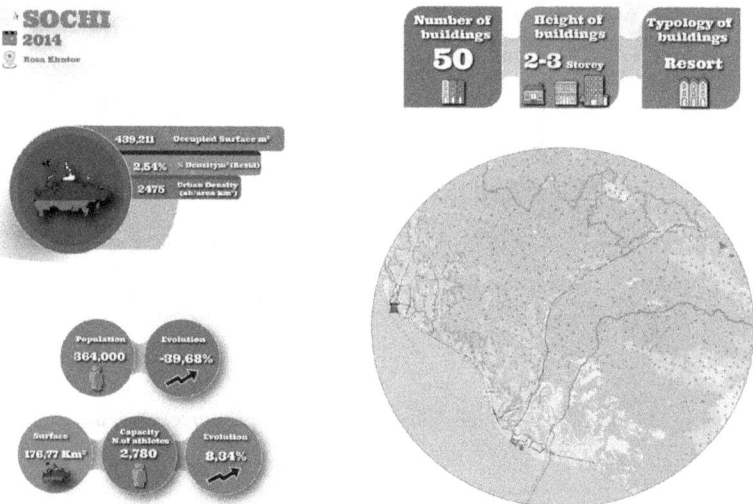

Figure 59. Model of the Olympic Village in Rosa Khutor 2014 (Source: Own implementation)

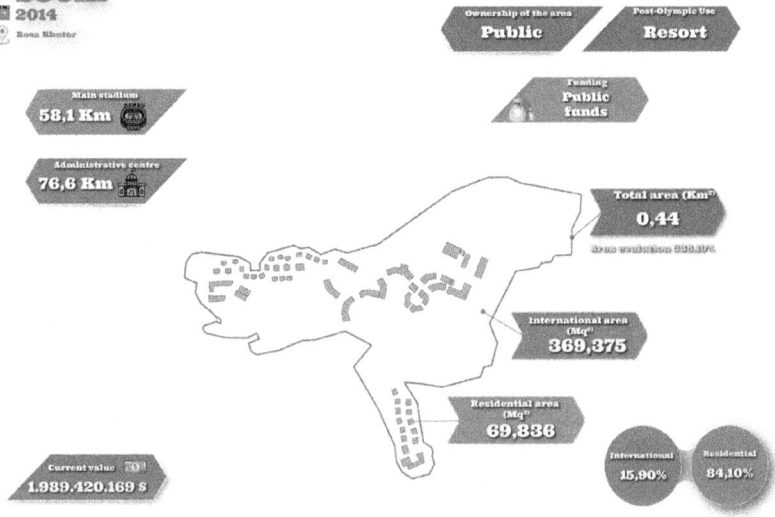

Figure 60. Model of the Olympic Village in Rosa Khutor 2014 (Source: Own implementation)

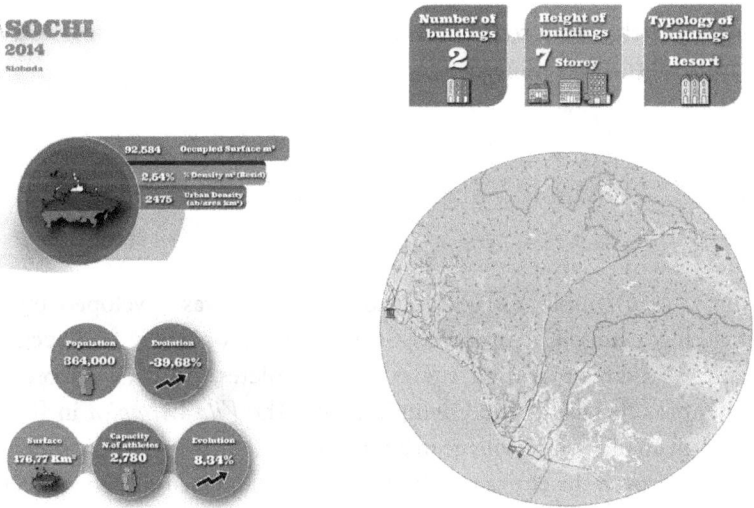

Figure 61. Model of the Olympic Village in Sloboda 2014 (Source: Own implementation)

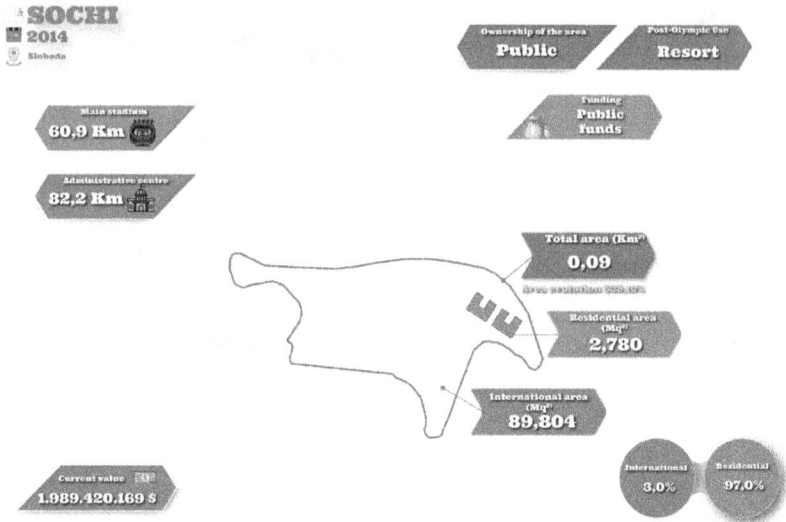

Figure 62. Model of the Olympic Village in Sloboda 2014 (Source: Own
implementation)

host cities that want to propose a mountain tourism project in a location
with a subtropical climate.

6.4. The Olympic Villages of Rio de Janeiro 2016

The 2016 Olympic Village in Rio de Janeiro was developed by constructing a huge Olympic Park in the suburb of Barra da Tijuca. The
entire settlement housed the coaches, athletes, doctors, referees and
Olympic officials for the Olympic event. The *Vila Olimpica* in Rio was
built on an old fishing village in an impoverished area of the capital city of
Rio. However, the entire area houses the training facilities, media centre,
stadiums and other services. The following project was realised following
the philosophy observed in the London 2012 edition.

6.4.1 *The Rio de Janeiro 2016 Olympic Village*

The Olympic Village comprises 31 buildings of 17 storeys and was designed to accommodate 18,000 people during and after the Olympic event. The buildings were designed by providing 3,604 accommodations and 10,160 bedrooms. The accommodations were designed to be resold as private flats in the post-Olympic period; however, to this day, the entire 75-hectare area continues to be abandoned. The initial plan was to convert the village into the Ilha Pura, a luxury housing complex where each flat was to be sold for an estimated price of $700,000.

The key locations of the Olympic event were connected by an internal shuttle service and a *rua carioca* footbridge that separated the residential area from the international Olympic Village Square. The village square was the central meeting place for the athletes and their guests, bringing together the different services and entertainment. In addition to the mandatory services, the Olympic Villa had a training centre that could accommodate almost 2,700 athletes per day. However, Rio in 2016 essentially

Figure 63. Location of the Olympic Villages in Rio de Janeiro 2016 (Source: Own implementation)

Figure 64. Rio de Janeiro, Olympic Village, current state (Source: CC)

turned into a major financial disaster. The cost of the games was financed by public and private money, and much of the infrastructure is abandoned or underused, including the $700 million Olympic Village that was to be transformed into luxury condominiums in the post-Olympic period. In 2016, it was estimated that only 7 per cent of the housing had been sold. However, after the following disaster, the city of Rio 2017 tried to sell the Olympic housing through low-interest financing for civil servants.

The following disaster reinforces the criticism of significant construction and real estate interests by the organisers, ignoring the problems induced by favelas in the capital of Rio.

In 2015, Mauricio Cruz Lopes, executive director of the real estate project, declared that the project was the largest in Brazil. However, the brand-new structure was in total contrast to the environment and the living conditions of the citizens. A few steps away from the Vila Autodromo Village, a favela had been demolished by 90 per cent to make way for the Olympic Park, and some families were also left without the provision of essential services such as water and electricity.

Figure 65. Rio de Janeiro, Olympic Village, current state (Source: CC)

The complex was built by a consortium of Odebrecht and Carvalho Hosken. Odebrecht is one of Brazil's big four contractors. Together with Andrade Gutierrez, Camargo Correa and OAS, Odebrecht built many of the projects for the World Cup and the Olympic Games.

In addition, the company was also involved in a corruption scandal involving the state-owned oil company Petrobras, triggering ongoing protests in cities across the country.

6.4.2 Rio de Janeiro 2016

A philosophy of urban expansion and infrastructural transformations in Rio de Janeiro strongly marked the Rio de Janeiro Olympic Village concept. Like the previous editions, the Rio edition had planned the Village close to the Olympic Park, where most venues were located. Services, roads, restaurants, transport and green spaces were built in this area. The

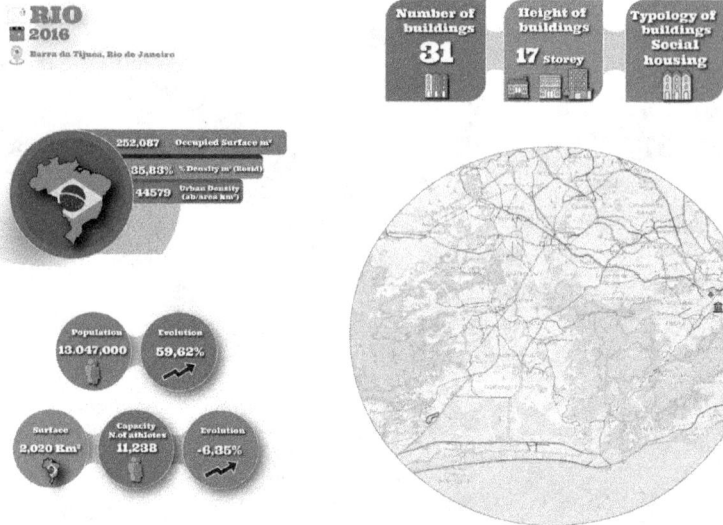

Figure 66. Model of the Olympic Village in Rio de Janeiro 2016 (Source: Own implementation)

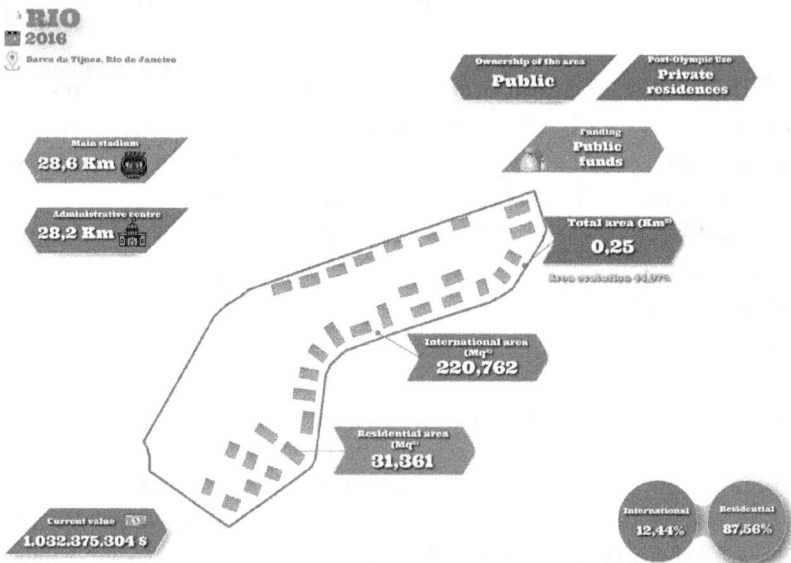

Figure 67. Model of the Olympic Village in Rio de Janeiro 2016 (Source: Own implementation)

accommodation capacity of the Olympic Village was for more than 18,000 athletes, and it was planned to be converted into a residence during the post-Olympic period[10]. The Rio Olympic Village project was very ambitious and still needs to find its function. The accommodation remains abandoned, and the entire Olympic area needs to be utilised more.

6.5. The Olympic Villages of Beijing 2022

The Beijing 2022 Olympic Villages will offer everyone an extraordinary Olympic experience, providing first-class services. Accommodations for athletes will be set up in Beijing, Yanqing and Zhangjiakou. The Beijing Village will be located south of the National Olympic Sports Centre, while the other two villages will be set up in mountain clusters in the centre of the Ice and Snow Cities. However, each of the three villages is located in the heart of each area and amid an ecological and sustainable environment. The Beijing Olympic project includes the removal of architectural barriers, and in the post-Olympic period, they will be put on the market. Therefore, the organising committee aimed to sell the Beijing and Zhangjiakou properties as residential properties. In contrast, the Yanqing property would be converted into a hotel with attached winter sports facilities. Finally, the state owns all land used to construct the Olympic Villages, and a governmental approach was adopted to ensure the project's quality and planned delivery.

All areas where the Olympic Villages were planned have abundant cultural and ecological resources. Therefore, the development of the Olympic Villages will not only benefit from these resources and help catalyse regional development and land resource management plans.

The design of the athletes' accommodation will adhere to the principles of environmental friendliness, safety, comfort, accessibility and

---

10   Organisation Committee of Rio de Janeiro 2016, Olympic and Paralympic Village: Rio 2016, Rio: Organizing Committee for the Olympic and Paralympic Games Rio 2016.

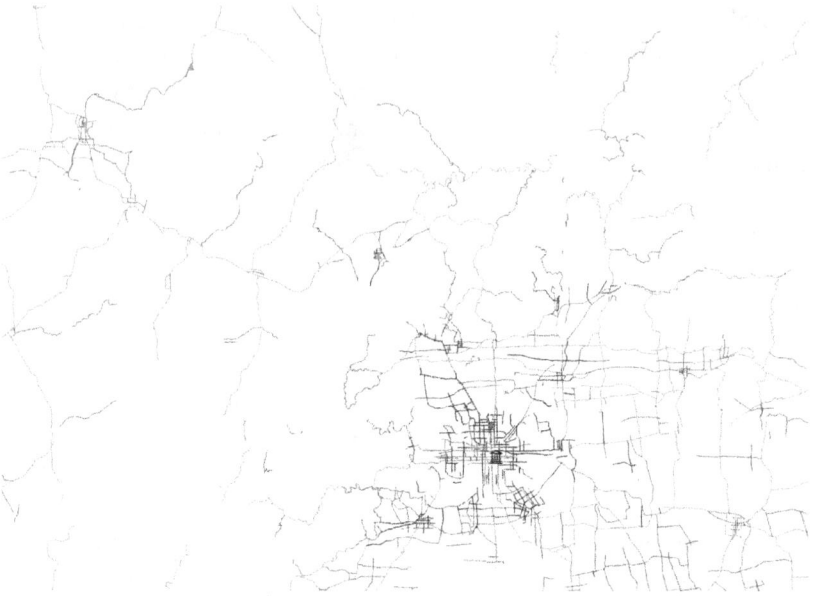

Figure 68. Location of the Olympic Villages in Beijing 2022 (Source: Own implementation)

sustainability. Therefore, following IOC sustainability requirements, the construction of each Olympic village will be conducive to the harmonious coexistence of humans and nature.

In addition, implementing energy and water-saving technologies, such as LED lighting, wastewater collection devices, and solar energy application systems, will allow for a low carbon footprint. In addition, greenhouse gas sensors have been installed in each building.

6.5.1 The Beijing 2022 Olympic Village

The Beijing Olympic Village was built south of the Olympic Green in a central area of the Chinese capital. In addition, the Village was built in an area adjoining the accommodation prepared for the 2008 summer event.

Just fifteen minutes away is the National Stadium, the site of the opening and closing ceremonies of the Games.

The Olympic Village was designed to fully use local advantages and highlight the appeal of local cultures, such as urban culture.

The entire Olympic accommodation area occupies 24.6 hectares and consists of three main zones: the Olympic Square (2.2 hectares), the Operations Zone (8.1 hectares), and the Residential Zone (14.3 hectares). In addition, the design includes Chinese cultural elements and is integrated into the Olympic Park by environmental sustainability criteria.

The entire area has different entrances, and all buildings and roads have been designed according to zero architectural barriers. In addition to housing, the entire residential area is provided with services, a multi-sports centre, a sports research centre, an outpatient clinic, a multi-faith centre, a fitness centre, a cultural centre and other ancillary facilities. To complete the area, a vast 13,000-m² shopping centre will be prepared.

Five hundred sixty flats will accommodate 2,260 athletes and Olympic officials within the residential area. After the event's end, the entire residential area will be transformed into a residential area like the Olympic Village of the 2008 summer edition. Meanwhile, the operational facilities will be transformed into recreational facilities for the community.

Figure 69. Beijing, Olympic Village, current state (Source: CC)

The development of the residential project has been included in the Beijing Ruban development plan. The Beijing Inno-Olympic Group Co. will be the state body responsible for developing the entire Olympic area. However, the Olympic Village is a permanent residential facility financed and built exclusively by the Beijing Inno-Olympic Group Co. In case of financial problems, the Beijing Municipal Bureau of Finance will oversee the underwriting.

6.5.2 The Yanqing 2022 Olympic Village

The Yanqing Olympic Village is located in the Xiaohaituo area of Yanqing County, with the Badaling section of the Great Wall to the southeast. In addition, the Yanqing Global Geopark is nearby, and it only takes five minutes for athletes to reach the National Gliding Centre and ten minutes to reach the National Alpine Skiing Centre.

In addition, the Olympic Village will utilise photovoltaic energy, geothermal energy, and hot springs for energy supply and health protection.

The entire accommodation area covers an area of 14.4 hectares at an altitude of 950 m. The Olympic Village consists of three main zones: the Olympic Square (0.6 hectares), the Operational Zone (6.6 hectares) and the Residential Zone (7.2 hectares). Complementing the Olympic area is a vast winter garden that runs through the residential zone and some outdoor recreational areas. Meanwhile, the ancillary facilities are the same as those prepared for the central Olympic Village in Beijing.

The residential area will consist of 760 flats, each with one or two single or double rooms, offering hotel accommodation for 1,430 athletes and officials.

After the Olympic event, Yanqing will be transformed into a tourist accommodation facility and a base for national ski training. Commercial activities, recreational services, catering facilities, and other services related to mountain tourism will provide the framework for the transformation of the ski resort. Therefore, the construction project aligns with Yanqing's Xiaohaituo ski development plan. At the same time, Beijing Enterprises Group Co will be the state-owned entity responsible for developing the

Figure 70. Yanqing, Olympic Village, current state (Source: CC)

entire residential area. However, the hotel will be financed and built exclusively by Beijing Enterprises Group Co. In case of financial problems, it will be supported by the Beijing Municipal Bureau of Finance.

6.5.3 *The Zhangjiakou 2022 Olympic Village*

The Zhangjiakou Olympic Village is in the centre of the Taizicheng Ice and Snow City, at the foot of the ancient Great Wall. In this village, all competition venues are within a 4-km radius and a five-minute drive from the Olympic Village.

In addition, the Zhangjiakou Olympic Village will benefit from local solar energy systems, wind power and local tourism resources.

The entire Zhangjiakou accommodation area covers an area of 21.9 hectares at an altitude of 1,600 m. The Olympic Village consists of three main zones: the Olympic Square (1.2 hectares), the Operational Zone (7.1 hectares) and the Residential Zone (13.6 hectares). The flats will be built within a forest so that all residents can enjoy the view of the nearby mountains. In addition, the Olympic Village will be equipped with the same functional areas as the Yanqing and Beijing villages.

The residential area will have 1,260 flats for 2,640 athletes and Olympic officials. In the period after the event, the entire area will become a vast residential complex for mountain tourism. Meanwhile, the operational facilities will be transformed into recreational facilities for the community.

The construction of the entire residential area is entirely in line with the Taizicheng Ice and Snow City Development Plan of Zhangjiakou. At the same time, Hebei Aoxue Investment Co. will be the state-owned entity responsible for the realisation and construction of the Olympic area. However, the permanent residential facility will be built and financed exclusively by the Beijing Enterprises Group Co. In case of financial problems, it will be supported by the Zhangjiakou Municipal Bureau of Finance.

Figure 71. Zhangjiakou, Olympic Village, current state (Source: CC)

As noted above, the Chinese project was developed with environmental compatibility and sustainable development principles. Therefore, according to the Organising Committee, the environmental impact will be minimised using appropriate construction materials and advanced environmental technologies. Among the implemented practices, we can identify the following actions.

- Sustainable construction materials are rapidly renewable with local resources,
- Alternative energy systems based on solar, wind and geothermal energy,
- Use of highly efficient mechanical and electrical systems,
- Innovative and ecological design,
- Wastewater reuse systems,
- Water-saving systems,
- Design of synthesised plastic recycling systems.

During the Olympic period, the Chinese capital's three Olympic villages will have 4,100 rooms and 6,330 beds, 30 per cent of which will be barrier-free. In addition, 1,430 beds will be reserved to accommodate the 4,900 athletes and officials.

6.5.4 *Beijing 2022*

The 2022 Beijing Olympics comprised three different Olympic areas in a territorial space that has reached a distance of 160 km between the central Olympic Village in the city and the Olympic Village in Yanqing. The Olympics would be held in three main hubs: Beijing, Yanqing and Zhangjiakou[11]. The construction of a central Olympic Village in the metropolis of Beijing was included in the area where the 2008 Summer Olympics were held. Meanwhile, the other Olympic Villages were

11   Beijing Organising Committee for the 2022 Olympic and Paralympic Winter Games, The Legacy Plan of the Olympic and Paralympic Winter Games Beijing 2022, Beijing, 2019.

planned in the two mountainous locations of Yanqing and Zhangjiakou. The Olympic Village in the city has 20 residential buildings and can accommodate 2,338 athletes during the Games. The Village was located in the centre of the metropolitan city, close to the National Olympic Sports Centre. During the post-Olympic phase, the project foresees that the residences will be rented from citizens and managed entirely by the central government. Meanwhile, the Olympic Villages in the mountain localities of Chaoyang and Yanging districts have some 3,500 beds. In contrast, the Zhangjiakou locality will be able to accommodate some 2,800 athletes and Olympic team officials.

The Yanging Olympic Village will predominantly host athletes competing in skating and alpine skiing. Meanwhile, the Zhangjiakou Village will host skiers, snowboarders and ski jumpers. The Yanging Olympic Village is at the foot of Haituo Mountain, surrounded by mountains and forests. The Village area is located in the centre of historical ruins in a diverse geological landscape and ecological environment. The Village consists of six groups of residences, which, in the post-Olympic period, will be transformed into hotels for the tourism development of Yanging County.

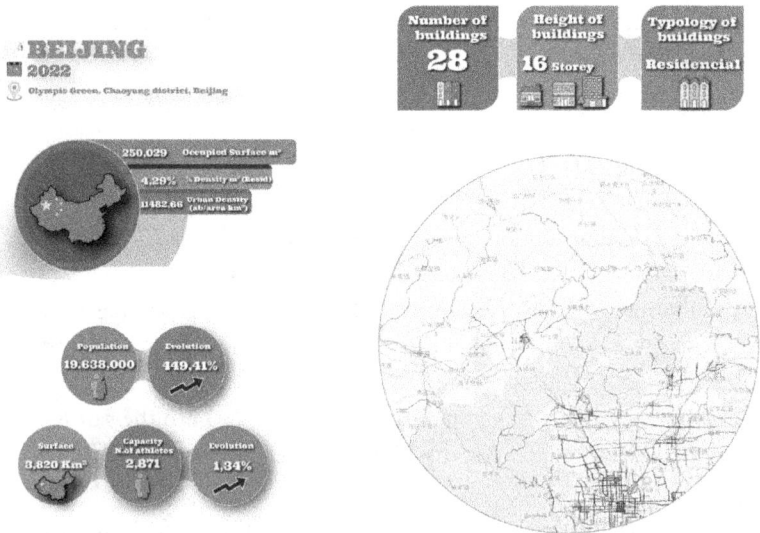

Figure 72. Model of the Olympic Village in Beijing 2022 (Source: Own implementation)

Meanwhile, Zhangjiakou Olympic Village is in the snow city of Taizicheng and occupies 197,600 m². The Olympic Village space comprises a residential and international area, including restaurants, clinics, gymnasiums, religious centres, leisure and training centres, residents' centres, delegation leader rooms, service centres and shared spaces. The residential area consists of thirty-one houses divided into ten groups, which will form a new mountain resort in the post-Olympic period to exploit winter tourism in Taizicheng.

The Beijing Games will be recognised as the most ambitious project ever in a regional Olympic area, which has never been seen before.

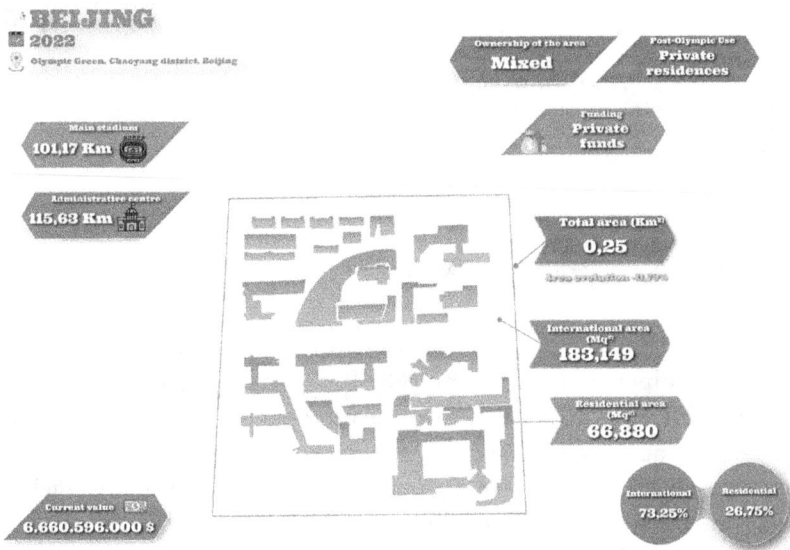

Figure 73. Model of the Olympic Village in Beijing 2022 (Source: Own implementation)

6.6. Considerations of the Summer Olympic Villages

As we have seen in the proposed comparative analyses, the Olympic Villages have evolved through different spatial models to solve housing problems at different historical stages. However, the projects in Helsinki, Rome, Mexico, Montreal, Seoul, Barcelona, Sydney, Beijing, London and Rio have inspired and continue to inspire host cities to propose permanent residential projects that respond to the demands of each specific metropolitan context. Olympic Villages have had different forms and different visions about housing. However, in recent years, the housing issue is less critical than it was after the Second World War. In recent years, we have consistently observed different solutions, including services and population levels, that could converge and become a new neighbourhood, encouraged by sports and cultural events. We will conclude by looking at each element's minimum and maximum values analysed to rank the Olympic Villages over time.

6.7. Considerations of the Winter Olympic Villages

As we have observed from the proposed comparative analyses, over the years, the Olympic Villages have evolved through different spatial models about the requirements of each specific location. However, the initial projects foresaw the use of hotels and resorts in places with a strong vocation for winter sports. From Oslo in 1952, the Olympic Village began to evolve through urban planning models that determined new accommodation constructions for the different cities. However, the Grenoble edition will mean the evolution of a regional model confirmed only with the Turin edition and the planning of three permanent Olympic Villages on a vast territory. However, in Calgary, in Salt Lake, we have observed a temporary accommodation solution that was converted into student accommodation. The Olympic Villages continue to inspire the different operating models of the new host cities. Over time, we have observed

Table 27. Comparative analyses of Summer Olympic Villages (Source: Own implementation)

Main stadium			
Min		Max	
Munich 1952	850 m.	Rio 2016	28,6 km.
Administrative centre			
Min		Max	
Barcelona 1992	3 km.	Rio 2016	28,2 km.
Number of buildings			
Min		Max	
Atlanta 1996	2	Sydney 2000	870
Urban density (ab/area kmq)			
Min		Max	
Berlin 1936	3.050,20 kmq	Atlanta 1996	126.677,39 kmq
% Density mq2 (residential)			
Min		Max	
Helsinki 1952	5,50%	Atlanta 1996	84,20%
Occupied area(mq2)			
Min		Max	
Atlanta 1996	81.451 mq2	Berlin 1936	1.299,260 mq2
Total Area (kmq)			
Min		Max	
Atlanta 1996	0,08 kmq	Berlin 1936	1,30 kmq
International area			
Min		Max	
Barcelona 1992	56,20%	Berlin 1936	97,31%
Residential area			
Min		Max	
Berlin 1936	2,69%	Barcelona 1992	43,80%
Current value			
Min		Max	
Los Angeles 1932	$8.217.404,43	Berlin 1936	$2.779.611.670,00

Table 28. Comparative analyses of Winter Olympic Villages (Source: Own implementation)

Main stadium			
Min		Min	
Grenoble 1968	600 m.	Grenoble 1968	101,17 km.
Administrative centre			
Min		Min	
Vancouver 2010	1,5 km.	Vancouver 2010	115,63 km.
Number of buildings			
Min		Min	
Squaw valley 1960 Innsbruck 1964	4	Squaw valley 1960 Innsbruck 1964	185
Urban density (ab/area kmq)			
Min		Min	
Calgary 1988	1.113,40 kmq	Calgary 1988	28.356,08 kmq
% Density mq2 (resid)			
Min		Min	
Grenoble 1968	1,76%	Grenoble 1968	21,80%
Occupied area(mq2)			
Min		Min	
Innsbruck 1964	38.475 mq2	Innsbruck 1964	1.278.087 mq2
Total Area (kmq)			
Min		Min	
Innsbruck 1964	0,03 kmq	Innsbruck 1964	1,27 kmq
Residential area			
Min		Min	
Innsbruck 1964	- 44,07%	Innsbruck 1964	98,75%
International area			
Min		Min	
Calgary 1988	1,25%	Calgary 1988	144,07%
Current value			
Min		Min	
Albertville 1992	$783.770,00	Albertville 1992	$6.660.596.000,00

different forms, typologies and structures that have become a heritage of the candidate cities or localities. However, the evolution of the number of athletes and Olympic competitions has increased the difficulty of organising the event through a territorial model supported by infrastructures. As noted above, since Turin 2006, the spatial dimension of the event has maintained a regional relevance through the main structure to be set up in the candidate metropolitan city. The Sochi, PyeongChang and Beijing editions confirm the trend of expanding the Olympic dimension on a regional territory. We will conclude by looking at each element's minimum and maximum values analysed to rank the Olympic Villages over time.

CHAPTER 7

Two Centuries of Olympic Urbanism

Based on the analyses conducted in the previous chapters and the theoretical framework provided by the following book, we can draw conclusions regarding the evolution of Olympic villages over time.

The project highlights some preliminary concepts fundamental to understanding Olympic urbanism: The Olympic Games and the Olympic Village.

The following concepts were consolidated through an in-depth analysis of the literature review and with the help of the results of della Sala's doctoral research.

The following research has provided valuable material for analysing and evaluating Olympic town planning during the twentieth and twenty-first centuries.

Therefore, within the following book, the mega-event has been considered a significant transformation process that, depending on the specific contexts, can manifest itself on the territory in different ways. As we observed in Chapter 3, territorial transformations through the organisation of a mega-event imply a physical and imaginary transformation of the host site. Moreover, among the most visible impacts over time, the following book highlights the importance of Olympic urbanism as a transformation process that implies rethinking the host cities' own strategies. A transformation process that, over time, has evolved to take on regional or transnational dimensions. Therefore, based on Olympic urbanism as a unitary corpus relating to the urban dimension of the sporting event, we highlight the Olympic Village as the only element capable of standing out over time as the fulcrum of urban development associated with the Olympic Games.

On the other hand, Olympic urbanism is part of a transformation process that manifests itself as a representation of different urban realities and in consideration of the specificities of each historical and socio-cultural context (della Sala, 2022).

Therefore, through the analysis of the Olympic Villages, we can relate the different projects using standard parameters that have helped us observe the evolution of Olympic accommodation during the twentieth and twenty-first centuries. Through the following contribution, we show how the Olympic Village is the central element in planning the Olympic event. Moreover, as noted above, in most cases, the Olympic Village has become a tangible asset and a common good for the host community. However, in observing the abandonments, the project must be articulated and designed by keeping the relationships this urban piece must establish with the existing urban fabric.

Therefore, we will conclude the following text with some reflections to provide an overview through a few fundamental points and in observance of the results obtained in the comprehensive study.

1. The Olympic Village as a catalyst for urban expansion and/or transformation

As noted in the previous chapters, Olympic Villages, both summer and winter, have taken on different forms that have evolved through other spatial models over time. Therefore, we can highlight an initial distinction between the various editions and events. Subsequently, given the evolution of spatial dimensions in the winter editions, the Olympic Village model has reached a territorial dimension that implies the construction of multiple lodgings dispersed over a regional territory. Moreover, the construction of numerous Olympic Villages, in the 2002 Beijing edition and more markedly in the future edition of Milan-Cortina 2026, will inevitably exceed 200 km in distance.

However, given the mountainous locations, creating new accommodations can catalyse the specific areas for tourist exploitation of the competition venues. In turn, an Olympic Village located in a metropolis can be considered a catalyst capable of fostering new urban transformations in neighbouring areas. Moreover, given the editions of Barcelona 1992, Turin 2006, Vancouver 2010, London 2012, Tokyo 2020 and Beijing 2022, the

Olympic Village planned and built in the central fabric of the cities favoured the birth of a new neighbourhood. Neighbourhoods in the post-event phase could favour the expansion of urban limits and imply a rethinking of transport and city mobility. However, given the accommodation size of the summer edition, the neighbourhood model of 20,000 people will have a totally different impact from the winter edition. In fact, the winter edition implies the construction of 5,000 accommodations. Housing can also be organised through other models or other types. Therefore, depending on the different forms, models of accommodation organisation, sizes, and distances, the Olympic Village can cause the transformation of an urban area. In addition, considering the areas of interest in each context, the Olympic Village can become a crucial element in expanding urban limits. Ultimately, the organisational model of the Olympic Village and its location brings about different forms of urban transformation that can take on other representations in terms of an urban artefact. Therefore, in light of the following element, the need arises to explain a fundamental aspect of the Olympic Village. Given the size of the project and the urban structure, the Olympic Village fits into urbanised places and must be integrated into the long-term planning of host cities. Only through integrated planning will it be possible to avoid drop-outs or overestimates in terms of housing size. Moreover, given the different urban, social, political, historical and architectural contexts, the Olympic Village can establish itself as a catalyst for multidimensional transformations within the core areas of our cities. Meanwhile, as far as the Olympic Village in the winter edition, the accommodation facilities built throughout history have always been characterised by a temporary character that, in the post-event phase, ensured a natural integration in the mountain areas. In addition, accommodation in mountain areas has allowed host communities to increase their accommodation stock and has become an asset available for the exploitation of sports tourism.

Thus, the Olympic Villages cannot be considered a catalyst for mountain communities' expansion or urban transformation. Meanwhile, for the summer edition, we can confirm that accommodation in central or peripheral areas can foster our cities' expansion and urban transformation.

2. The Olympic legacy as a tool for analysing the transformation of
   host cities

By the analyses conducted and considering the concepts of mega-events
and impact, at an academic level, studies have mainly focused on a purely
economic analysis or one closely related to physical urban infrastructure.
The following evolution of literary studies needs to pay more attention to
the importance of the legacy of heritage through urban transformations.
As noted in the second and third chapters, during the Olympic history,
projects were transformed in consideration of the needs and priorities
of the host cities. Moreover, since Rome in 1960, Olympic projects have
become a tool for restructuring, redesigning and rethinking host city
strategies.

   Therefore, the Olympic event cannot only be observed through a purely
economic analysis. Moreover, given urban planning as a long-term legacy,
the analysis of the event in the post-Olympic period, using observation of
the territory, will favour the analysis of the impact and the different forms
it can take in the reference territory.

   For this reason, the legacy of heritage is of primary importance in
evaluating Olympic transformations. In addition, the existing heritage
should be considered in the preliminary assessment of the new urban artefact. Furthermore, as we noted in Chapter 2, heritage legacy represents a
unique opportunity for the renewal of candidate cities. Given the urban
transformation process, urban heritage is essential for introducing new
practices about the conservation and exploitation of available heritage.

   However, considering the results obtained within the qualitative inter-
view (See della Sala, 2022), in the period following the Turin 2006 Olympic
event, the city and the mountain sites could enjoy and exploit the legacy
of the available heritage. Therefore, we can consider the Olympic event as
an artefact that can relate to the territory and take different forms in territorial systems over time. In addition, considering the relationships that
the Olympic heritage will assume, new areas and territorial dynamics will
develop in the post-Olympic period. Consequently, the consideration of
Olympic urbanism with the help of purely economic elements does not
guarantee the observation of the intangible values of the legacy over time.

Therefore, the Olympic Village should be considered as an inheritance that, over time, can be viewed as an asset available for implementing new strategies in the host cities.

3. The planning models of the Olympic Villages built during the twentieth and twenty-first centuries

In view of the observations made in the second and third chapters, we have identified some standard spatial models that have allowed an initial classification of the different Olympic experiences. Therefore, during the twentieth and twenty-first centuries, Olympic Villages were planned through other spatial models identified as monocentric, satellite, polycentric, cluster, peripheral and metropolitan. Therefore, the following models allow us to distinguish between the models proposed for the summer and winter editions.

In the summer edition, the Olympic Villages took different forms, highlighting the monocentric model as a standard model capable of recognising the existence of an urban strategy. Furthermore, since the Barcelona edition 1992, the monocentric model has been characterised as the new planning model for Olympic accommodation within the modern metropolis.

Therefore, by identifying the following spatial models, we can state that the different experiences of the Olympic Villages in the summer edition took on a dimension that was included within the central strategies of the host metropolises.

On the contrary, in the winter edition, the Olympic Villages were planned with the help of different models. In the winter edition, as noted in Chapter 3, the polycentric model can be identified as a standard model that has affirmed the existence of a regional strategy since Turin 2006. This strategy has distinguished itself as the new planning model for multiple Olympic Villages in an ever-expanding Olympic space. Therefore, the increasing size of the winter edition allowed us to observe how the polycentric model catalysed regionalisation processes within the host territories.

Therefore, common patterns can be identified in both Olympic editions. Spatial models allow us to classify and observe the different experiences each Olympic Village model assumed during the event's planning and post-event phase through the spatial dimension. Furthermore, given the theoretical framework, we highlight the importance of the following models as inspiring examples for future event planning. Therefore, the spatial model of Olympic receptivity can establish different relationships that, in each local context, tend to produce acts of territorialisation by creating connections and interconnections that will transform existing networks. On the other hand, winter spatial organisation implies different relationships with different places in the broader area.

For this reason, the planning of the winter edition will be more complex from a local development point of view. A regional spatial pattern may contribute to the creation of different spatial imbalances. Imbalances that may foment the creation of internal competition between intermediate areas, which in the future may turn into territories excluded from spatial planning.

4. The Urban Functions of the Olympic Villages in the Post-
   Olympic Period

Considering the urban functions assumed by the Olympic Villages, we can observe different urban strategies linked throughout history to the housing needs of the host cities. Therefore, the construction of the Olympic Village assumed other urban functions that can be analysed by observing its uses: temporary or permanent.

However, a clear distinction between the two editions can be observed through the classification according to the character of the Olympic Villages.

In the summer edition, except for the first and American editions, the permanent accommodation model was the most used. Meanwhile, in the winter edition, temporary accommodation was only used in locations with hotels or a project to enhance tourism in the areas affected by the Olympic event.

Subsequently, looking at the post-event evolution, the Olympic Villages can be analysed regarding reuse and abandonment of accommodation. In the summer edition, as noted in Chapter 3, most Olympic

Villages were reused as available accommodation for the host community. Furthermore, in the summer edition, only the Olympic Villages of Berlin 1936, Athens 2004 and Rio de Janeiro 2016 are abandoned and disused by the local community. On the other hand, in the Winter edition, Olympic accommodation was mainly implemented through the use or construction of hotels, resorts or tourist accommodations.

Thus, in the winter edition of mountain resorts, Olympic Villages were added or included in tourist circuits, becoming an asset available for winter tourism exploitation in the post-event phase. Including accommodation within the receptivity did not compromise the territory and thus reduced the abandonment phenomena related to the Olympic facilities.

Moreover, in the winter edition, except for the Sarajevo bombing, the Olympic Village of Turin 2006 was the only abandoned accommodation facility. Therefore, observing the urban functions of the Olympic Villages in the post-event phase, it is stated that the accommodation in the mountain sites was always reused. However, in the summer edition, and considering the different sizes, different abandonments were observed in the post-event phase. Furthermore, the abandonment of sites can catalyse other problems and forms of social deviance, such as squatting. The classification of Olympic Villages advanced in the following text allows us to consider the importance of post-Olympic planning. Planning must consider the model, use and size that the accommodation complexes assume in a phase of uncertainty. Therefore, in view of the reuse of the Olympic Village, it is argued that a specific or mixed function must be planned before the construction phase.

5. The Olympic Village and Strategic Planning

Regarding the following aspect, it must be emphasised that the cultural, social, political, and historical context of the host cities is decisive in the strategic planning of the Olympic Village. However, given the theoretical framework and the qualitative interviews on the case of Turin 2006 (See della Sala, 2022), the master plan is highlighted as a fundamental element for the revaluation and reconstruction of the urban fabric of industrial cities.

Furthermore, the importance of the master plan in reducing the possibilities of real estate speculation in areas that in the post-event phase, could be transformed through the acquisition of other meanings and functions is highlighted. Subsequently, the city's strategic plan will establish new guidelines for exploiting the master plan. Therefore, the future development of each specific context can be realised through the procedures. Furthermore, it must be emphasised that the Olympic Village must inevitably be included in the city's strategic plan to be integrated into the urban fabric. Only through the inclusion of housing in a strategic plan will it be possible to define different patterns of future urban redevelopment.

Thus, the identification of the Olympic Village area must consider the obligations of the master plan and the future objectives of the strategic plans of the metropolises themselves. However, the strategic plan increases the number of goals, strategies, and action plans. Moreover, the strategic plan allows for greater control of the overall strategic management. In this specific case, it is necessary to emphasise participatory planning as a fundamental process for designing future Olympic accommodations. Only through co-participation can the integration of the Olympic Villa within the area's social, cultural and entrepreneurial fabric be guaranteed. Thus, preparing a strategic plan will allow for a common strategy for the city and its citizens.

Through the following points, we have put forward different hypotheses that converge on considering the winter edition as the model that has reached a spatial dimension that will inevitably provoke a rethinking of regional strategies. Finally, the following hypotheses will allow us to consider how the new elements of legacy and sustainability since the Turin 2006 edition have become critical elements for the assignment of the Olympic edition.

However, in the specific case of Turin 2006, the interviewees emphasised that the Olympic Villages at the mountain sites were built as a physical and available capital for the entire community. Therefore, the importance of the capacity of the Olympic Village and the available accommodation is stated.

I.   Development and Validation of Olympic Urbanism

Observing the different urban planning experiences of the organising cities of the summer and winter Olympic events during the twentieth and twenty-first centuries, the results reveal the existence of a unified corpus related to the urban planning dimension of the Olympic Games. The unitary corpus comprises three physical elements: the athletics stadium, the Olympic pool facilities, and the "Olympic Village". The following elements must be considered as the main characteristic elements of Olympic urbanism. Furthermore, as can be seen from the analysis of conventional urban development in host cities, the Olympic event has been a catalyst for other urban processes throughout history. During the twentieth and twenty-first century urban development of cities in the post-war phase, one observes an Olympic event embedded within host cities' reconstruction, expansion and transformation processes.

Moreover, given the different urban programmes of the host cities, the physical dimension, architectural style and spatial pattern of the Olympic Village constitute the central elements of Olympic urbanism. However, the development of Olympic urbanism in the winter and summer editions has evolved completely differently. An evolution that has allowed an alternative form of urban planning to emerge. A form that places the Olympic Village at the centre of urban transformations. Moreover, given the event's evolution, the Olympic Village has established itself as a practice centre for new accommodation within the host cities. Accommodations that will be available for the post-Olympic period and that can foster the implementation of good practices within the host community.

II.   The Olympic Village as an Urban Strategy for the Host City

As noted above, the evolution of Olympic urbanism during the twentieth century in the summer edition clearly shows a consolidation of urban strategies. Urban strategies in which the construction of the Olympic Village became a physically localised element in the central areas of the candidate cities. Thus, the Olympic Village became a constitutive element of long-term urban transformation strategies.

Meanwhile, the evolution of twenty-first-century Olympic town planning in the winter edition clearly shows the consolidation of a new spatial dimension within regional strategies.

Within this regional dimension, we can observe how the Olympic Village has become a multiple element in different locations within the Olympic space.

In this sense, Olympic Villages in cities are no longer temporary intensive-use places that become a territorial unknown in the post-Olympic period. On the other hand, Olympic Villages in mountain sites, in observation of the Turin 2006 case, have become part of a tourism strategy for consolidating the new post-Olympic territorial promotion model. On the contrary, Olympic lodgings in the centre of host cities are constituted as building blocks that are part of the recent long-term urban change and transformation programmes. As the cases of Barcelona 1992, Turin 2006, London 2012 and Tokyo 2020 clearly demonstrate, Olympic accommodations have become multidimensional urban catalysts. Therefore, implementing the city's strategic plan will favour the inclusion of the Olympic Village in a consolidated urban strategy. An urban strategy that, regarding revaluation and redefinition of areas, will implement the regions and strategies included in the long-term plan. On the other hand, in the winter edition, the Olympic Village, through its inclusion in a territorial promotion strategy, can become an element that produces value and multiple results for the mountain sites.

However, in editions such as Sochi 2014 or PyeongChang 2018, we observed a spatial dimension that proposed the transformation of new territories for the exploitation of sports tourism.

Both cases allow us to verify the existence of a new strategy of tourism exploitation in recent locations, which can become a risk in terms of the production of territory and non-use of accommodation facilities. Thus, the Winter Olympic Village can only be a strategy to increase tourist accommodation in places that already connect with winter sports.

III. The recent rethinking of the Olympic legacy

Throughout the twentieth-century Olympic experience, the defining element of Olympic town planning has been the construction of physical infrastructure. This infrastructure has become increasingly complex with a new set of intangible elements associated with the urban legacy of the Games (Olympic legacy). Since the Games of Barcelona 1992, Sydney 2000, Turin 2006, Beijing 2008 and especially in the case of London 2012, Olympic urbanism has become a fundamental element in fostering new values of social and cultural identity associated with the collective legacy of the event. Therefore, the transformation of host cities with the help of the new sustainability criteria represents a new era for the urban legacy of the Olympic event. Applying new criteria in constructing the Olympic Village and new sports facilities means an intangible possibility for host cities. The consideration of existing facilities and the creation of temporary spaces has introduced new elements into Olympic design. Furthermore, introducing new urban recycling elements can ensure the reuse of urban facilities in the post-Olympic period and promote further intangible transformation processes in host cities. Again, given the new selection criteria for candidate cities, legacy and sustainability have become two critical elements for the Olympic bid evaluation process. Therefore, the future projects of Paris 2024, Los Angeles 2028, and Brisbane 2032 allow us to observe that the elements of sustainability and urban legacy have been introduced since the beginning of the Olympic bid. Furthermore, including the following elements in the bid can be considered critical components for awarding the Olympic event. Therefore, considering the Olympic legacy for planning the Olympic Village as an optional event is a fundamental tool for assessing the urban legacy of the host cities. In conclusion, the Olympic legacy must take on specific planning for the post-event phase. In this sense, the Olympic legacy has become the most crucial element in achieving a winning strategy regarding event allocation. However, it can be concluded by emphasising the need to plan a post-Olympic strategy with a reach of at least ten to fifteen years after the Olympic event.

IV. The Winter Olympic Games and their evolution towards the
     development of metropolitan and regional strategies

By analysing the different urban planning experiences of the cities that
hosted the Winter Olympics during the twentieth and twenty-first centuries, we can determine a new evolution of the spatial dimension and
urban planning models. As noted in Chapter 2, the winter edition developed through primordial models that were not essentially urban
planning, initially associated with the idea of a mountain resort and a
temporary use of the sporting event. However, over time, starting with
the 1952 Oslo edition, the Olympic event began to develop through
more complex and substantial urbanisation models. Similarly, as noted
in Chapter 3, the Olympic Village will show the exact evolution of construction criteria observed in the summer edition. The endpoint of the
urbanisation process will be the proposal of metropolitan and regional
strategies in the Olympic experiences following Turin 2006. Therefore,
the territorial model observed in the Turin 2006 edition helps us to establish the validity of a central structure based on the metropolitan city
and other venues distributed across the regional territory. Thus, the new
territorial expansion model of the winter edition is configured in different locations, creating and reforming networks in a broader restricted
space that is constantly expanding. The following model allows us to
observe how the mountain resorts and metropolitan cities will be the
central elements of the new territorial promotion strategy. In addition,
mountain resorts will be prominent in holding sports competitions. The
latest development model will manifest itself as the maximum expression
of the true extent of the metropolisation phenomenon of the host cities.
However, developing a regional strategy should consider mountain areas
and the city as two projects.

     Given the Turin 2006 event, mountains were excluded from the overall
project in the post-Olympic phase. Thus, it is emphasised that the Winter
Olympics can affect the territory through different scales of intervention.
Moreover, regionalisation and temporal connections dissolve and disappear
in the post-Olympic period. However, the Milan-Cortina 2026 edition, in
which three regions and two major cities will participate, will again change

the dimension of the Olympic space. The establishment of a central city, a cluster in the mountains and other competition venues distributed in a new spatial dimension will exponentially increase the surface area of the Olympic area until a maximum distance of 370 km is reached between the city of Milan and Anterselva di Mezzo, the competition venue.

Finally, it is necessary to emphasise the importance of strategic planning in the perspective of sustainable development of future territories. Integrating strategic lines in a joint development framework can reduce the possibilities of failure and uncertainty observed in the post-event phase. Furthermore, it must be said that the Olympic Games are special events that must fit into the ordinary planning of the host cities and territories. Therefore, the Olympic Village must be planned through a long-term strategy to become a dynamic instrument for the host territory. For this reason, the Olympic Village has become an essential part of the debate and research on living within the host cities. In observance of the future projects of Paris 2024, Milan-Cortina 2026, Los Angeles 2028 or Brisbane 2032, we see how the Olympic Village has become the only urban element that will be planned in observance of the construction priority of new housing. Therefore, participatory planning and local authorities' collaboration are paramount for implementing lines and strategies in terms of sustainable development. In conclusion, the following book offers new evaluation criteria for Olympic Villages and contemporary elements to investigate for a more equitable and sustainable future. For this reason, given the interviews carried out (See della Sala, 2022), the importance of organising a new Olympic symposium to discuss, investigate and observe the different Olympic experiences that could not be kept at that time is emphasised.

Furthermore, the organisation of a new symposium on Olympic villages could encourage the consideration of new construction, evaluation, and integration criteria aligned with the UN's Sustainable Development Goals. Finally, it is argued that the Olympic Village must be identified in a strategic planning context, investigating the relationship between the city and the Olympic Village. This is a nexus that needs to be addressed by most previous studies. The relationship and relationship between the city and the Olympic accommodation must be considered as a basis for future studies and research projects. Therefore, the following book opens the door to further research for studying and evaluating past or future Olympic Villages through standard parameters. Thus, it emphasises the importance of the longitudinal perspective introduced by the study (See della Sala, 2022), which can also be applied to past events. In particular, the following book on Olympic Villages and its relationship to urbanisation must be considered an original contribution to the existing literature and the first significant study on Olympic urbanism.

Bibliography

Andranovich, G., Burbank, M. J., & Heying, C. H. (2001). Olympic cities: Lessons learned from mega-event politics. *Journal of Urban Affairs, 23*(2), 113–131.

Andreff, W. (2012). The winner's Curse: Why is the cost of mega sporting events so often underestimated? In W. Maenning (Ed.), *International handbook on the economics of mega sporting events*. Elgar.

Andreff, W. (2013). Economic development as major determinant of Olympic medal wins: Predicting performances of Russian and Chinese teams at Sochi Games. *International Journal of Economic Policy in Emerging Economies, 6*(4), 314–340.

Athens 2004 Organising Committee for the Olympic Games; Efharis Skarveli, I. Z. (Ed.). (2004). *Official report of the XXVIII Olympiad: Athens 2004*.

ATHOC. (2005). *Official Report of the XXVIII Olympiad*, 2 vols. Athens: Liberis Publications Group.

Auge, M. (1992). *LOS «NO LUGARES» ESPACIOS DEL ANONIMATO Una antropología de la Sobremodernidad*. Edition de Seuil.

Batuhan, T. (1996). *The politics of Olympic transportation planning: The case of the 1996 26 Summer Olympic Games*, pp. 68–72.

Batuhan, T. (2012). The politics of Olympic transportation planning: The case of the 1996 Atlanta Summer Olympic Games. *Problems, possibilities, promising practices: Critical dialogues on the Olympic and Paralympic Games*, 68.

Bauman, Z. (1998). *Globalization: The human consequence*. Cambridge: Polity Press.

Beijing Organizing Committee for the Games of the XXIX Olympiad. (2010). *Official report of the Beijing 2008 Olympic Games*.

Billings, S. B., & Holladay, J. S. (2012). Should cities go for the gold? The long-term impacts of hosting the Olympics. *Economic Inquiry, 50*(3), 754–772.

Bischel, T., Gatzeva, M., Hambrock, M., Kwan, B., & Lim, C. (2011). *Olympic Games Impact (OGI) study for the 2010 Olympic and Paralympic Winter Games games-time report the Vancouver Organizing Committee for the 2010 Olympic and Paralympic Games (VANOC)*.

Blunden, H. (2012). The Olympic Games and housing. In *The Palgrave handbook of Olympic studies* (pp. 520–532). Palgrave Macmillan UK.

Bohigas, O. (2003). *Realismo, urbanidad y fracasos*. Servicio Publicaciones ETSA.

Bondonio, P. (2007). *A giochi fatti : le eredità di Torino 2006*, p. 374.

Bondonio, P. E. D. (2006). *Olimpiadi, oltre il 2006: Torino 2006: secondo rapporto sui territori olimpici*. Carocci.

Bondonio, P., & Campaniello, N. (2006). Torino 2006: What kind of Olympic Winter Games were they? A preliminary account from an organizational and economic perspective. *Olympika: The International Journal of Olympic Studies, 15*, 1–34.

Bondonio, P. V., Dansero, E., & Mela, A. (2006). *Olimpiadi, oltre il 2006: Torino 2006, secondo rapporto sui territori olimpici, 337*, p. 358.

Bosia, D., & Savio, L. (2016). From the management of the project, to the evidence of the results: the Olympic Village of Turin 2006. *TECHNE-Journal of Technology for Architecture and Environment, 12*, 137–143.

Bourdieu, P. (1978). Sport and social class. *Social Science Information, 17*(6), 819–840.

Brenner, N. (1998). Global cities, glocal states: Global city formation and state territorial restructuring in contemporary Europe. *Review of International Political Economy, 5*(1), 1–37.

Brenner, N. (2002). *Space of neoliberalism, urban restructuring in north America and western Europe.* Blackwell.

British Olympic Council. (1909). *The fourth Olympiad: Being the official report of the Olympic Games of 1908 celebrated in London.*

Brogan, P. (1996). Cities and the 'World events' process. *Town & Country Planning, 65*(11): 314–316.

Brunet, F. (1995). An economic analysis of the Barcelona '92 Olympic Games: Resources, financing, and impact. *The Keys to Success: The Social, Economic and Communications Impact of* Barcelona '92, 203–237.

Brunet, F. (2005). *The economic impact of the Barcelona Olympic Games, Barcelona: The legacy of the games 1992–2002* (pp. 1–27). Barcelona: Centre d'Estudis Olímpics (UAB).

Burns, J. A. and Mules, T. J. (1986). A framework for the analysis of major special events. In J. A. Burns, J. H. Hatch and T. J. Mules (eds), *The Adelaide Grand Prix*, pp. 5–36. Adelaide: The Centre for South Australian Economic Studies.

Cashman, R. (1999). *Staging the Olympics* (Centre for Olympic studies). UNSW Press.

Cashman, R. (2002). What is "Olympic Legacy"?. In *The legacy of the Olympic Games.* Losanna: International Olympic Committee (IOC).

Cashman, R. (2005). *The bitter-sweet awaking: The legacy of Sydney 2000 Olympic Games.* Sidney: Walla Walla Press.

Cashman, R. (2010). *Impact of the games on Olympic host cities.* Centre d'Estudis Olímpics (UAB).

Cashman, R., & Hughes, A. (1998). *The green games: A golden* opportunity (Center of Olympic studies). Sydney: University of New South Galles.

Chalip, L. (2010). *Leveraging the Sydney Olympics for tourism.* Centre d'Estudis Olímpics, UAB.

Chalip, L. H., & McGuirty, J. (2004). Bundling sport events with the host destination. *Journal of Sport & Tourism, 9*(3), 267–282.

Chalkley, B., & Essex, S. (1999). Urban development through hosting international events: A history of the Olympic Games. *Planning Perspectives, 14*(4), 369–394.

Chappelet, J. L. (2002). From Lake Placid to Salt Lake City: The challenging growth of the Olympic Winter Games since 1980. *European Journal of Sport Science, 2*(3), 1–21.

Chappelet, J. L. (2008). *The international Olympic committee and the Olympic system: The govern.* Routledge.

Chappelet, J.-L. (2008). Olympic environmental concerns as a legacy of the Winter Games. *The International Journal of the History of Sport, 25*(14), 1884–1902.

Chappelet, J.-L. (2010). *A short overview of the Olympic Winter Games.* CEO-UAB.

Chappelet, J.-L. (2010). Mega sporting event legacies: A multifaceted concept. *Papeles de Europa, 25,* 76–86.

Chappelet, J. L. (2016). From Olympic administration to Olympic governance. *Sport in Society, 19*(6), 739–751.

Chappelet, J., & Junod, T. (2006). A tale of 3 Olympic cities: What can Turin learn from the Olympic legacy of other Alpine cities. *Major sport events as opportunity for development. Valencia: Valencia summit proceedings,* 1–8.

Charlesworth, E. R. (Ed.). (2005). *Cityedge: Case studies in contemporary urbanism.* Routledge.

Chen, Y., Qu, L., & Spaans, M. (2013). Framing the long-term impact of mega-event strategies on the development of Olympic host cities. *Planning Practice and Research, 28*(3), 340–359.

CHORE. (2001). *The impacts of the Sydney Olympic Games on housing rights the impacts of the Sydney Olympic Games on housing rights 2.* Geneva: Centre of Housing Rights and Evictions.

CHORE. (2007). *Multi-stakeholder guidelines on mega-events and the protection and promotion of housing rights.* Geneva: Centre of Housing Rights and Evictions.

Clark, G. (2008). *Local development benefits from staging global events: Achieving the local development legacy from 2012, (2011),* 89. LEED.

Clarke, E., & Gaile, G. (1998). *The work of cities.* Minneapolis: University of Minnesota Press.

COJO 76. (1978). *Montreal 1976: Games of the XXI Olympiad Montreal 1976: Official report.*

Collins, A., Jones, C., & Munday, M. (2009). Assessing the environmental impacts of mega sporting events: Two options? *Tourism Management, 30*(6), 828–837.

Comitato Giorgio Rota. (2000). *Rapporti su Torino – Lavori in corso.*

Comitato Giorgio Rota. (2001). *Rapporti su Torino – La mappa del mutamento.*

Comitato Giorgio Rota. (2002). *Rapporti su Torino – Voglia di cambiare.*

Comitato Giorgio Rota. (2003). *Rapporti su Torino – Count down.*

Comitato Giorgio Rota. (2004). *Rapporti su Torino – Le radici del nuovo futuro.*

Comitato Giorgio Rota. (2005). *Rapporti su Torino – L'immagine del cambiamento.*

Comitato Giorgio Rota. (2006). *Rapporti su Torino – Giochi aperti.*

Comitato Giorgio Rota. (2007). *Rapporti su Torino – Senza rete.*

Comitato Giorgio Rota. (2008). *Rapporti su Torino – Solista e solitaria.*

Comitato Giorgio Rota. (2009). *Rapporti su Torino – 10 anni per un'altra Torino.*

Comitato olimpico nazionale italiano. (1957). *VII Giochi olimpici invernali, Cortina d'Ampezzo, 1956 = VII Olympic Winter Games*, Cortina d'Ampezzo, 1956.

Comitato per l'Organizzazione dei XX Giochi Olimpici Invernali Torino 2006. (2007). *XX Giochi Olimpici invernali Torino 2006 = XX Olympic Winter Games Torino 2006.*

Comité d'organisation des Jeux Olympiques et Paralympiques d'hiver de 2010 à Vancouver. (2010). *VANOC official Games report = Rapport officiel des Jeux COVAN.*

Comité d'organisation des Xèmes Jeux Olympiques d'hiver. (1969). *Rapport officiel [Xèmes Jeux olympiques d'hiver] : Official report [Xth Winter Olympic Games].*

Comité d'organisation des XVes Jeux Olympiques d'hiver. Calgary. (1988). *Rapport officiel des XVes Jeux Olympiques d'hiver = XV Olympic Winter Games official report.*

Comité de candidature de Beijing aux Jeux Olympiques d'hiver de 2022. (2014). Beijing 2022: Candidate city.

Comité exécutif des jeux d'Anvers ; Comité olympique belge. (1920). XVIIème Olympiade Anvers 1920.

Comité Olímpico Brasileiro. (2010). *Passion unites us: Rio 2016 bid official report = Unis par la passion: rapport officiel de la candidature de Rio 2016.*

Comité olympique français. (1924). *Les Jeux de la VIIIe Olympiade: Paris 1924: rapport officiel.* Comité olympique français.

Comité Olympique Suisse. (1951). *Rapport général sur les Ves Jeux Olympiques d'hiver, St-Moritz 1948.*

Comité organisateur des Jeux de la XIX olympiade. (1969). Mexico 68: Official report.

COOB'92. (1992). *Official report of the games of the XXV Olympiad Barcelona 1992.*

Cook, I. G. (2006). Beijing as an 'internationalized metropolis'. In: Wu, F. (Ed.) *Globalisation and China's cities* (pp. 63–84). London: Routledge.

Currie, G. and Shalaby, A. (2012). Synthesis of transport planning approaches for the world's largest events. *Transport Reviews*, 32, 113–136.

Decreto Legge. (2003). *Torino 2006 – La legge n. 48 del 2003.*

Dansero, E. (2002). I "luoghi comuni" dei grandi eventi. Allestendo il palcoscenico territoriale per Torino 2006. *Bollettino della Società Geografica Italiana, 7,* 861–894.

Dansero, E. (2014). I grandi eventi: spazi per una discreta geografia del cambiamento continuo. *Atti del XXXI Congresso Geografico Italiano*, vol. II. Mimesis.

Dansero, E., & De Leonardis, D. (2006). Torino 2006. La territorializzazione olimpica e la sfida dell'eredità. *Bollettino della società geografica italiana,* 611–641.

Dansero, E., Maroni, O., & Ricciardi, C. (2003). Cercando SLOT per le Valli Chisone e Germanasca. In *Una geografia dei luoghi per lo sviluppo locale. Approcci metodologici e studi di caso, Slot quaderno 3,* 111–145. Baskerville.

Dansero, E., & Mela, A. (2004). Trasformazioni territoriali e ambientali come eredità di Torino 2006. Le percezioni degli attori del territorio olimpico. In *Aspettando le Olimpiadi. Primo rapporto sui territori olimpici* (pp. 109–153). Carocci.

Dansero, E., & Mela, A. (2007). Olympic territorialization. The case of Torino 2006. *Journal of Alpine Research / Revue de géographie alpine, 95–3,* 16–26.

Dansero, E., & Mela, A. (2012). Bringing the mountains into the city: Legacy of the Winter Olympics, Turin 2006. In *The Palgrave handbook of Olympic studies* (pp. 178–194). Palgrave Macmillan UK.

Dansero, E., & Puttilli, M. (2010). Mega-events tourism legacies: The case of the Torino 2006 winter Olympic Games – a territorialisation approach. *Leisure Studies, 29*(3), 321–341.

Davidson, M., & McNeill, D. (2012). The redevelopment of Olympic sites: Examining the legacy of Sydney Olympic Park. *Urban Studies, 49*(8), 1625–1641.

Davis, J. (2015). Urban design on deprivation: Exploring the role of Olympic legacy framework masterplanning in addressing spatial and social divides. In *Mega-event cities: Urban legacies of global sports events*. Routledge.

DCLG. (2015). *London 2012 Olympics regeneration legacy evaluation framework.*

DCMS. (2012a). *Department for culture, media and sport London 2012 Olympic and Paralympic games impacts and legacy evaluation framework final report.*

DCMS. (2012b). *Plans for the legacy from the 2012 Olympic and Paralympic games.*

DCMS. (2015). *Olympic Games impact study-London 2012 post-games report.*

Debord, G. (1992). The society of the spectacle. 1967. *Paris: Les Éditions Gallimard.*

De Carlo, M., D'angella, F., De, M., & Phd, C. (2003). *Strategic repositioning of heritage destinations through large-scale events management destination management and the management of large-scale events in cultural destinations.*

De Pieri, F., & F. Giulietta. (2008). *The Olympic sites: The architecture of 2006 in the urban landscape of Turin.* Torino: Umberto Allemandi.

De Rossi, A., & Durbiano, G. (2006). *Turin 1980–2011: Its transformation and its images*, 119.

della Sala, V. (2022a). The Olympic Village and the Olympic urbanism: Perception and expectations of Olympic specialists. *Bollettino della Società Geografica Italiana serie 14, 5* (2), 51–64.

della Sala, V. (2022b). *The Olympic Villages and Olympic urban planning: Analysis and evaluation of the impact on territorial and urban planning (XX–XX I centuries)*. Doctoral thesis. UAB. POLITO.

della Sala, V. (2023a). Olympic Games and expectations: The factor analysis model about Olympic Urbanism and Olympic Villages. *Sociologia e Ricerca Sociale, 132*(3), 127–147.

della Sala, V. (2023b). Sustainable planning for the Olympic legacy. *Environmental Analysis & Ecology Studies, 11* (2), 1237–1239.

della Sala, V. (2023b). Olympic Games: Between expectations and fears. Factor analysis model applied to Olympic urbanism and Olympic Villages. *Rivista Internazionale di Scienze Sociali, 132*(1), 55–86.

della Sala, V. (2023c). The legacy of the Turin 2006 Olympic Games through a long-term development perspective. Reflection and opinion about the physical and social change in the post-Olympic period. *Cambio. Rivista sulle trasformazioni sociali, 25* (1), 229–247.

Deloitte. (2012). *Lessons from London 2012*.

Delorme, F. (2014). Du Village-station à la station-Village. Un siècle d'urbanisme en montagne. *In Situ. Revue des patrimoines, 24* <https://journals.openedition.org/insitu/11243>.

Dematteis, G. (1995). *Progetto implicito. Il contributo della geografia umana alle scienze del territorio*. Milano: Franco Angeli.

Dematteis, G., & Governa, F. (a cura di) (2005). *Territorialità, sviluppo locale, sostenibilità. Il modello SLoT*. Milano: Franco Angeli.

Dematteis, G., & Guarrasi, V. (1995). *Urban networks, Heo-Italy, Italian committee for international geographical union*. Bologna: Patron Editore. Vol. 2-1994.

Dickson, G., & Schofield, G. (2005). Globalisation and globesity: The impact of the 2008 Beijing Olympics on China. *International Journal of Sport Management and Marketing, 1*(1–2), 169–179.

Essex, S. J., & De Groot, J. (2016). The winter Olympics: Driving urban change, 1924–2022. In *Olympic Cities* (pp. 84–109). Routledge.

Essex, S., & Chalkley, B. (1998). Olympic Games: Catalyst of urban change. *Leisure Studies, 17*(3), 187–206.

Essex, S., & Chalkley, B. (2004). Mega-sporting events in urban and regional policy: A history of the Winter Olympics. *Planning Perspectives, 19*(2), 201–204.

Essex, S., & Chalkley, B. (2010). *Urban transformation from hosting the Olympic Games.* Centre d'Estudis Olímpics (UAB).

European Commission. (2011). *Cities of tomorrow.* European Commission.

European Commission. (2016). *The state of European cities 2016. Cities leading the way to a better future.* European Commission.

Evans, G. (1995). *The role of the festival in urban regeneration: Planning for the British millennium festival.* In *International Festivals Association* Second European Research Symposium, Edinburgh (Vol. 17).

Evans, G. (2009). Creative cities, creative spaces and urban policy. *Urban Studies, 46*(5&6), 1003–1040.

Feddersen, A., Maennig, W., & Zimmermann, P. (2008). How to win the Olympic Games: The empirics of key success factors of Olympic bids. *Revue d'économie Politique, 118*(2), 171–187.

Ferrari, S. (2002). *Event marketing: i grandi eventi e gli eventi speciali come strumenti di marketing.* Padova: CEDAM.

Flyvbjerg, B., & Stewart, A. (2012). Olympic proportions: Cost and cost overrun at the Olympics 1960–2012. *Social Science Research Network. SSRN Electronic Journal,* (June), 1–23.

Frawley, S., & Adair, D. (2013). *Managing the Olympics: Managing the Olympics.* Palgrave Macmillan.

Furrer, P. (2002). Giochi Olimpici sostenibili: utopia o realtà?. *Bollettino della Società Geografica Italiana, serie XII, VII,* 4.

Garcia, B. (2005). Deconstructing the City of Culture: The Long-term Cultural Legacies of Glasgow 1990. *Urban Studies, 42*(5–6), 841–868.

Gehl, J. (2006). *La humanización del espacio urbano: la vida social entre los edificios – Dialnet.* Reverté.

Georgiadis, K., & Theodorikakos, P. (2016). The Olympic Games of Athens: 10 years later. *Sport in Society, 19*(6), 817–827.

Getz, D. (1989). Special events: Defining the product. *Tourism Management, 10*(2), 125–137.

Getz, D. (1991) *Festivals, special events and tourism.* New York: Van Nostrand Reinhold.

Getz, D. (1997). Trends and issues in sport event tourism. *Tourism Recreation Research, 22*(2), 61–62.

Getz, D. (2004). Bidding on events. *Journal of Convention & Exhibition Management, 5*(2), 1–24.

Getz, D., & Page, S. J. (2016). *Event studies: Theory, research and policy for planned events* (3rd edn). Routledge.

Getz, D., & Fairley, S. (2008). Media management at sport events for destination promotion: Case studies and concepts. *Event Management, 8*(3), 127–139.

Girginov, V., & Hills, L. (2008). A sustainable sports legacy: Creating a link between the London Olympics and sports participation. *The International Journal of the History of Sport, 25*(14), 2091–2116.

Gold, J., & Gold, M. M. (Eds) (2016). *Olympic cities: City agendas, planning, and the World's Games, 1896* (3rd edn). Routledge.

Gold, J. R., & Gold, M. M. (2008). Olympic cities: Regeneration, city rebranding and changing urban agendas. *Geography Compass, 2*(1), 300–318.

Gordon, B. F. (1983). *Olympic architecture: Building for the Summer Games.* New York: Wiley.

Governa, F. (1999). *Il milieu urbano. L'identità territoriale nei processi di sviluppo.* Milano: Franco Angeli.

Graham, S., & Marvin, S. (2002). *Splintering urbanism: Networked infrastructures, technological mobilities and the urban condition.* Routledge.

Gratton, C., & Henry, I. (2001). *Sport in the city: Sport in the city.* Routledge.

Gratton, C., & Preuss, H. (2008). Maximizing Olympic impacts by building up legacies. *The International Journal of the History of Sport, 25*(14), 1922–1938.

Gratton, C., Shibli, S., & Coleman, R. (2005). Sport and economic regeneration in cities. *Urban Studies, 42*(5–6), 985–999.

Greslery, G. (1994). La Ciutat Mundial in *Visions Urbanes. Europa 1870-1993. La ciutat de l'artista. La ciutat de l'arquitecte.* Barcelona : Centre de Cultura Contemporània de Barcelona, Electa, 164–166.

GRI. (2014). *Sustainability reporting guidelines, 2015(17/5).*

Guala, C. (2002). *International society of city and regional planners 38th congress – "The Pulsar Effect" monitoring Torino 2006 Winter Olympic Games.*

Guala, C. (2002). Per una tipologia dei mega eventi. In E. Dansero & A. Segre (a cura di), *Il territorio dei grandi eventi. Riflessioni e ricerche guardando a Torino 2006* (Vol. VII, pp. 743–755). Roma: Bollettino della società geografica italiana.

Guala, C. (2015). *Mega eventi: immagini e legacy dalle Olimpiadi alle Expo.* Carocci.

Guala, C., & Crivello, S. (2006). Mega events and urban regeneration: The background and numbers behind Turin 2006. In N. Muller, M. Messing, & H. Preuss (a cura di), *From Chamonix to Turin: The Winter Games in the scope of Olympic research* (pp. 323–342). Kassel: Agon.

Guarrasi, V. (2002). Ground zero: Grandi eventi e trasformazione urbane. *Bollettino della Società Geografica Italiana. serie XII,* VII, 4.

Hall, C. M. (1989). The definition and analysis of hallmark tourist events. *Geojournal: Geography of Tourism and Recreation, 19*(3), 263–268.

Hall, C. M. (1992). *Hallmark tourist events: Impacts, management & planning.* London: Belhaven Press.

Hall, P. (1984). *The world cities.* Widenfeld & Nicolson.

Harvey, D. (1991). *The condition of postmodernity: An enquiry into the origins of cultural change*. Wiley.

Harvey, D. (2002). *Spaces of capital: Spaces of capital*. Routledge.

Harvey, D. (2012). *Rebel cities: From the right to the city to the urban revolution*. Verso.

Haugen, H. Ø. (2005). Time and space in Beijing's Olympic bid. *Norsk Geografisk Tidsskrift, 59*(3), 217–227.

Heine, M. (2018). Olympic commodification and civic spaces at the 2010 winter Olympic Games: A political topology of contestation. *International Journal of the History of Sport, 35*(9), 898–910.

Hensher, D. and Brewer, A. (2002). Going for gold at the Sydney Olympics: How did transport perform? *Transport Reviews, 22*, pp. 381–399.

Hiller, H. (1984). Toward a science of Olympic outcomes: The urban legacy. *The legacy of the Olympic Games*, 2000, 102–109.

Hiller, H. (2014). *Host cities and the Olympics: An interactionist approach* (1st edn). Routledge.

Hiller, H. H. (1990). The urban transformation of a landmark event. *Urban Affairs Quarterly, 26*(1), 118–137.

Hiller, H. H. (1998). Assessing the impact of mega-events: A linkage model. *Current Issues in Tourism, 1*(1), 47–57.

Hiller, H. H. (2000a). Mega-events, urban boosterism and growth strategies: An analysis of the objectives and legitimations of the Cape Town 2004 Olympic bid. *International Journal of Urban and Regional Research, 24*(2), 449–458.

Hiller, H. H. (2000b). Toward an urban sociology of mega-events. *Research in Urban Sociology, 5*, 181–205.

Hiller, H. H. (2003). Towards a science of Olympic outcomes: The urban legacy. In H. Hiller (Ed.), *The legacy of the Olympic Games* (pp. 32–38). Losanna: International Olympic Committee.

Hodges, J., & Hall, C. (1996). The housing and social impacts of mega events: Lessons for the Sydney 2000 Olympics. In G. Kearsley (Ed.), *Tourism down under II; towards a more sustainable tourism* (pp. 152–166). Dunedin: Centre for Tourism, University of Otago.

Holcomb, B. & Parisi, P. (1990). Press and place: Equivocal relations in the redeveloping city. Center for Urban Policy Research, Rutgers University, Research Paper no. 20. Piscataway, New Jersey.

Holden, M., Mackenzie, J., & VanWynsberghe, R. (2008). Sports mega- events: Social scientific analyses of a global phenomenon. *The Sociological Review*. New York: Blackwell Publishing.

HOLSA. (1992). Barcelona Olimpica. La ciudad Renovada, La gran transformacion urbana de Barcelona. Ambit Serveis Editorials.

Horne, J., & Manzenreiter, W. (2006). An introduction to the sociology of sports mega-events. *The Sociological Review, 54*(2_suppl), 1–24.

Hughes, H. (1993). Olympic tourism and urban regeneration; 1996 Summer Olympics. *Festival Management & Event Tourism, 1*(4), 137–184.

Imrie, R., Lees, L., & Raco, M. (2008). *Regenerating London: Governance, sustainability and community in a global city.* Routledge Taylor & Francis Group.

Insolera, I. (1978). *La città e la crisi del capitalismo.* Tempi Nuovi Laterza.

IOC. (1996). *Technical Manual on Olympic Village.*

IOC. (2005a). *International Olympic committee international Olympic committee manual for candidate cities for the games of the xxix Olympiad 2008.*

IOC. (2005b). *Technical manual on Olympic Village (November 2005).*

IOC. (2005c). *Technical manual on planning, coordination & management of the Olympic Games (November 2005).*

IOC. (2005d). *Technical manual on sport (November 2005).*

IOC. (2005e). *Technical manual on venues – design standards for competition venues (November 2005).*

IOC. (2007). *Technical manual on Olympic Games impact.*

IOC. (2008). *Manual for candidate cities for the games of the xxix Olympiad 2008.*

IOC. (2010a). *Candidature procedure and questionnaire xxiii Olympic Winter Games.*

IOC. (2010b). Olympic Games: Legacies and impacts/Jeux Olympiques: Héritages et Impacts. *Legacy* (December), 1–64.

IOC. (2011a). *Candidature acceptance procedure games of the xxxii Olympiad.*

IOC. (2011b). *Internationalism in the Olympic movement.* VS Verlag für Sozialwissenschaften.

IOC. (2012). *Olympic legacy.*

IOC. (2013a). *Factsheet: Legacies of the games* (December), 1–11.

IOC. (2013b). *The games of the Olympiad* (October), 1–8.

IOC. (2014). *Factsheet the Olympic Winter Games.*

IOC. (2015a). *Factsheet legacies of the games.*

IOC. (2015b). *Olympic Winter Games strategic review working group.*

IOC. (2015c). *Part II: Candidature file.*

IOC. (2015d). *Technical manual on Olympic Village (November 2005).*

IOC. (2016). *Key OGKM activities.*

IOC. (2017a). *Celebrate capture partner embed legacy strategic approach moving forward.*

IOC. (2017b). *Host city contract operational requirements.*

IOC. (2017c). *Host city contract principles games of the xxxiii Olympiad in 2024.*

IOC. (2017d). *Report evaluation commission 2024.*

IOC. (2017e). *The Olympic Winter Games in numbers: Vancouver 2010, Sochi 2014 and PyeongChang 2018.*

IOC. (2018a). Host city contract operational requirements. Retrieved from <https://stillmed.olympics.com/media/Document%20Library/OlympicOrg/Games/Host-City-Contract/HCC-Operational-Requirements.pdf>

IOC. (2018b). *Olympic agenda 2020 Olympic Games: The new norm report by the executive steering committee for Olympic Games delivery.*

IOC. (2018c). *Olympic summer games villages from Paris 1924 to Rio 2016.*

IOC. (2018d). *Sharing history, enriching the future Olympic Winter Games villages from Oslo 1952 to PyeongChang 2018.*

IOC. (2019). *IOC sustainability progress update a review of our 2020 objectives.*

IOC. (2020). *Olympic charter.*

Jacobs, J. (1961). *The death and life of great American cities.* New York: Random House.

Jennings, W. (2012). *Olympic risks.* Palgrave Macmillan UK.

Jia, H., Lu, Y., Shaw, L. Y., & Chen, Y. (2012). Planning of LID-BMPs for urban runoff control: The case of Beijing Olympic Village. In *Separation and purification technology, 84,* 112–119. Elsevier.

Kagaya, S. (1991). Infrastructural facilities provision for Sapporo's winter Olympic of 1972 and its effects on regional developments. *Revue de Géographie Alpine, 79*(3), 59–71.

Karamichas, J. (2012). The Olympics and the environment. In Helen Jefferson Lenskyj & Stephen Wagg (Eds), *The Palgrave handbook of Olympic studies* (pp. 381–393). Palgrave Macmillan UK.

Kasimati, E. (2003). Economic aspects and the Summer Olympics: A review of related research. *International Journal of Tourism Research, 5*(6), 433–444.

Kasimati, E. (2006). *Macroeconomic and financial analysis of mega-events: Evidence from Greece,* p. 249.

Kaspar, R. (April–May 1998). *Sport environment and culture, Olympic review: Official publication of the Olympic movement* (No. XXVI-20, pp. 67–70). Lausanne: COI.

Kassens-Noor, E. (2013). Transport legacy of the Olympic Games, 1992–2012. *Journal of Urban Affairs, 35*(4), 393–416.

Kassens-Noor, E. (2016). From ephemeral planning to permanent urbanism: An urban planning theory of mega-events. *Urban Planning, 1*(1), 41–54.

Kassens-Noor, E., & Lauermann, J. (2017). How to bid better for the Olympics: A participatory mega-event planning strategy for local legacies. *Journal of the American Planning Association, 83*(4), 335–345.

Kearns, G., & Philo, C. (1993). *Selling places: The city as cultural capital, past and present (Policy planning and critical theory)* (1st edn). Pergamon Press.

Kidd, B. (1996). *The Toronto Olympic commitment: Towards a social contract for the Olympic Games.* International Symposium Lausanne. IOC.

Koolhaas, R. (1995). *The generic city.* Sikkens Foundation.

Kotler, P. (1993). *Marketing places, attracting investment, industry, and tourism to cities, states, and nations.* The Free Press.

Kovac, I. (2002). *The Olympic territory: A way to an ideal Olympic scene, in the legacy of the Olympic Games 1984–2000.* International Symposium Lausanne. IOC.

Larson, M. (2009). Joint event production in the jungle, the park, and the garden: Metaphors of event networks. *Tourism Management, 30*(3), 393–399.

Larson, M., Getz, D., & Pastras, P. (2015). The legitimacy of festivals and their stakeholders: Concepts and propositions. *Event Management, 19*(2), 159–174.

Le Corbusier. (1942). *Principio De Urbanismo.* Paris: Ariel.

Le Corbusier. (1965). *Maniera Di Pensare l'urbanistica.* Bari: Universale Laterza.

Lefebvre, H. (1969). *El derecho a la ciudad.* Ediciones penisula.

Lefebvre, H. (1991a). *Critique of everyday life.* London: Verso.

Lefebvre, H. (1991b). *The production of space | Wiley.* Wiley.

Lenskyj, H. J. (2006). The Olympic (affordable) housing legacy and society responsibility. In *Eighth international symposium for Olympic research* (pp. 191–199).

Lenskyj, H. J. (2012). The winter Olympics: Geography is destiny? In *The Palgrave handbook of Olympic studies* (pp. 88–102). Palgrave Macmillan UK.

Lenskyj, H. J., & Wagg, S. (2012). *The Palgrave handbook of Olympic studies.* Palgrave Macmillan.

Leopkey, B., & Parent, M. M. (2012). Olympic Games legacy: From general benefits to sustainable long-term legacy. *International Journal of the History of Sport, 29*(6), 924–943.

Liao, H., & Pitts, A. (2006). A brief historical review of Olympic urbanization. *The International Journal of the History of Sport, 23*(7), 1232–1252.

Logan, J. R., & Molotch, H. L. (1987). *Urban fortunes, the political economy of place.* University of California Press.

The London Organising Committee of the Olympic Games and Paralympic Games. (2013). *London 2012 Olympic Games: The official report.*

Lynch, K. (1960). Kevin Lynch: The image of the city 1. *The Image of the City,* 1–14.

Macintosh, D., & Hawes, M. (1992). The COI and the world of interdependence. *Olympika, 1,* 29–45.

Maenning, W., & Zimbalist, A. (2012). *International handbook on the economics of mega sporting events.* Elgar.

Magnaghi, A. (2000). *Progetto locale.* Torino: Bollati Boringhieri.

Magnaghi, A. (2001). *Rappresentare i luoghi.* Firenze: Alinea Editrice.

Masterman, G. (2009). Strategic sports event management: Olympic edition.

Metropolis. (2002). *Commission 1 the impact of major events on the development of large cities.*

Milano Cortina 2026 Candidate City Olympic Winter Games. (2019). Milano Cortina 2026/Milano Cortina 2026 Candidate City Olympic Winter Games.

Millet i Serra, L. (1997). *Olympic Villages after the games Lluís Millet.* Centre d'Estudis Olímpics, UAB.

Modrey, E. M. (2008). Architecture as a mode of self-representation at the Olympic Games in Rome (1960) and Munich (1972). *European Review of History: Revue Europeenne d'histoire, 15*(6), 691–706.

Montanari, A. (2002). Grandi Eventi, marketing urbano e realizzazione di nuovi spazi turistici. *Bollettino della Società Geografica Italiana. serie XII, VII,* 4.

Moragas, M. (1996). *Olympic Villages: A hundred years of urban planning and shared experiences: International symposium on Olympic Villages* (Autonomous University of Barcelona, Ed.). Lausanne.

Moragas, M. De, Botella, M., & Lausanne, O. M. (1995). *The keys to success: The social sporting, economic and communications impact of Barcelona '92.*

Müller, M. (2011). State dirigisme in megaprojects: Governing the 2014 winter Olympics in Sochi. *Environment and Planning A: Economy and Space, 43*(9), 2091–2108.

Müller, M. (2014). After Sochi 2014: Costs and impacts of Russia's Olympic Games. *Eurasian Geography and Economics, 55*(6), 628–655.

Muñoz, F. (1996). Historic evolution and urban planning typology of Olympic Village. In *Hundred years of urban planning and shared experiences, International symposium on Olympic Villages.* IOC.

Muñoz, F. (2006). Olympic urbanism and Olympic Villages: Planning strategies in Olympic host cities, London 1908 to London 2012. *The Sociological Review, 54*(2_suppl), 175–187.

Muñoz, F. (2008). Urbanización: Paisajes Comunes. *Lugares Globales.*

Muñoz, F. (2011). I grandi eventi nella città del XXI secolo: variazioni sull'esperienza di Barcellona. *Rivista di sociologia urbana e rurale, 96,* 46–71.

Muñoz, F. (2015). Urbanalisation and city mega-events: From "copy&paste" urbanism to urban creativity. In Valerie En Viehoff & Gavin Poynter (Eds), *Mega-event cities: Urban legacies of global sports events* (pp. 11–21). Farnham, Surrey: Ashgate.

Nakamura, H., & Suzuki, N. (2017). Reinterpreting Olympic legacies: The emergent process of long-term post-event strategic planning of Hakuba after the 1998 Nagano winter games. *International Journal of Sport Policy, 9*(2), 311–330.

Netherlands Olympic Committee (Committee 1928); Ed. by G. Van Rossem; transl. by Sydney W. Fleming (1928). *The ninth Olympiad: Being the official report of the Olympic Games of 1928 celebrated at Amsterdam.*

Oben, T. (2011). *Sport and the environment: An UNEP perspective* (pp. 25–33).

Odoni, A., Stamatopoulos, M., Kassens, E. & Metsovitis, I. (2009). Preparing an airport for the Olympic Games: Athens. *Journal of Infrastructure Systems, 15,* 50–59.

OECD. (2020). *Cities in the world* (OECD urban studies). OECD.

OECD. (2011). *OECD regional outlook 2011*. OECD.

Olds, K. (1998). Urban mega-events, evictions and housing rights: The Canadian case. *Current Issues in Tourism, 1*(1), 2–46.

Olympic Winter Games Committee Lake Placid. (1932). *III Olympic Winter Games, Lake Placid 1932: Official report.*

Organisasjonskomiteen. (1953). *VI Olympiske Vinterleker Oslo 1952 = VI Olympic Winter Games Oslo 1952.*

Organisationskomitee der IX. Olympischen Winterspiele in Innsbruck 1964. (1967). *Offizieller Bericht der IX. Olympischen Winterspiele Innsbruck 1964.*

Organisationskomitee für die IV. Olympischen Winterspiele 1936 Garmisch-Partenkirchen. (1936). *IV Olympische Winterspiele 1936: Garmisch-Partenkirchen 6. bis 16. Februar : amtlicher Bericht .*

Organisationskomitee für die XI. Olympiade Berlin 1936. (1937). *The XIth Olympic Games Berlin, 1936: Official report.*

Organising Committee. (1955). *The official report of the organising committee for the games of the XV Olympiad.*

Organising Committee for the XIV Olympiad. (1951). *The official report of the organising committee for the XIV Olympiad.*

Organising Committee of the Games of the XXII Olympiad. (1981). *Games of the XXII Olympiad: Official report.*

Organising Committee of the XIVth Winter Olympic Games 1984 at Sarajevo. (1984). Final report = Rapport final = Završni izvještaj.

Organizing Committee. (1928). *Rapport général du Comité exécutif des IImes Jeux olympiques d'hiver et documents officiels divers.*

Organizing Committee. (1958). *The official report of the organizing committee for the games of the XVI Olympiad Melbourne 1956.*

Organizing Committee. (1960). *VIII Olympic Winter Games Squaw Valley, California, 1960: Final report California Olympic Commission.*

Organizing Committee for the Games of the XVII Olympiad. (1963). *The Games of the XVII Olympiad Rome, 1960: the official report of the Organizing Committee.*

Organizing Committee. (1966). *The Games of the XVIII Olympiad, Tokyo 1964: The official report of the organizing committee.*

Organizing Committee. (1974). *Die Spiele: The official report of the organizing committee for the Games of the XXth Olympiad Munich 1972.*

Organizing Committee. (1995). *Official report of the XVII Olympic Winter Games Lillehammer 1994.*

Organizing Committee. (1996). *Official Report of the Games of the XXVI Olympiad.* Atlanta, GA: Atlanta Committee for the Olympic Games.

Organizing Committee for the Games of the XVII Olympiad. (1963). *The games of the XVII Olympiad Rome, 1960: The official report of the organizing committee.*

Organizing Committee for the XIth Olympic Winter Games Sapporo 1972. (1973). *The XI Olympic Winter Games Sapporo 1972: Official report = Les XI Jeux olympiques d'hiver Sapporo 1972: rapport officiel.*

Organizing Committee for the XIIth Winter Olympic Games 1976 at Innsbruck. (1976). *Endbericht : XII. Olympische Winterspiele Innsbruck 1976 = Rapport final: Innsbruck '76 = Final report: Innsbruck '76.*

The Organizing Committee for the XVIII Olympic Winter Games Nagano 1998. (1999). *The XVIII Olympic Winter Games: Official report Nagano 1998.*

Organizing Committee of Albertville. (1992). *Rapport officiel des XVIes Jeux Olympiques d'hiver d'Albertville et de la Savoie = Official report of the XVI Olympic Winter Games of Albertville and Savoie.* Olympic Winter Games. Organizing Committee.

Organizing Committee of Lake Placid 1980. (1980). *XIII Olympic Winter Games Lake Placid 1980: Official results = XIII Olympic Winter Games Lake Placid 1980: résultats officiels = XIII Olympic Winter Games Lake Placid 1980: offizielle Ergebnisse.*

Organizing Committee of Los Angeles Olympic. (1985). *Official report of the Games of the XXIIIrd Olympiad Los Angeles 1984.*

Organizing Committee of the XXII Olympic Winter Games and XI Paralympic Winter Games of 2014 in Sochi (2015).

Owen, K. A. (2002). The Sydney 2000 Olympics and urban entrepreneurialism: Local variations in urban governance. *Australian Geographical Studies, 40*(3), 323–336.

Paquette, J., Stevens, J., & Mallen, C. (2011). The interpretation of environmental sustainability by The International Olympic Committee and Organizing Committees of the Olympic Games from 1994 to 2008. *Sport in Society, 14*(3), 355–369. London: Taylor And Francis.

Paris Candidate City Olympic Games 2024. (2016). *Candidature file: Paris candidate city Olympic Games 2024.*

Payne, M. (2007). *A gold-medal partnership.* London: Strategy+Business. PWC.

Pentifallo, C. (2013). *A case study approach to indicator-based impact assessment: The Olympic Games impact (OGI) study and the Vancouver Olympic athletes' village in contextual perspective.* IOC.

Pentifallo, C., & VanWynsberghe, R. (2015). Mega-event impact assessment and policy attribution: Embedded case study, social housing, and the 2010 Winter Olympic Games. *Journal of Policy Research in Tourism, Leisure and Events, 7*(3), 266–281.

Poynter, G. (2009). *London assembly economic development, culture, sport and tourism committee literature review: Olympic legacy governance arrangements professor Gavin Poynter November 2009*, (November), pp. 1–47.

Poynter, G. (2010). *Mega events and the urban economy: What can Olympic cities learn from each other?* Centre d'Estudis Olímpics (UAB).

Poynter, G. (2012). The Olympics: East London's renewal and legacy. In *The Palgrave handbook of Olympic studies* (pp. 505–519). Palgrave Macmillan UK.

Poynter, G. (2016). Urbanism: Space, planning and place-making. In *Mega-event cities: Urban legacies of global sports events* (pp. 59–61). Routledge.

Preuss, H. (1998). Problemizing arguments of the opponents of Olympic Games. In *Global and cultural critique: Problematizing the Olympic Games: 4th international symposium for Olympic research (January 1998)*, pp. 197–218.

Preuss, H. (2000). *Economics of the Olympic Games: Hosting the games 1972–2000*. Walla Walla Press, University of Germany.

Preuss, H. (2004). *The economics of staging the Olympics: A comparison of the games 1972–2008*. New York: Edward Elgar Publishing.

Preuss, H. (2007). The conceptualisation and measurement of mega sport event legacies. *Journal of Sport and Tourism*, *12*(3–4), 207–228.

Preuss, H. (2010). *The economic dimension of the Olympic Games*. Centre d'Estudis Olímpics (UAB).

Preuss, H., & Hong, S. P. (2021). Olympic legacy: Status of research. *Journal of Global Sport Management*, *6*(3), 205–211.

Preuss, H., Andreff, W., & Weitzmann, M. (2019). *Cost and revenue overruns of the Olympic Games 2000–2018: Cost and revenue overruns of the Olympic Games 2000–2018*. Springer Fachmedien Wiesbaden.

Public Health Act. (1925). United Kingdom.

PyeongChang Organising Committee for the XXIII Olympic and Paralympic Winter Games. (2019). *PyeongChang 2018 official report.*

Raco, M. (2013). Governance as legacy: Project management, the Olympic Games and the creation of a London model. *International Journal of Urban Sustainable Development*, *5*(2), 172–173. Taylor & Francis.

Raffestin, C. (1981). *Per una geografia del potere*. Milano: Unicopli.

Raffestin, C. (1984). Territorializzazione, deterritorializzazione, riterriteritorializzazione e informazione. In A. Turco (a cura di), *Regione e regionalizzazione* (pp. 69–82). Milano: Franco Angeli.

Regione Piemonte. (2006). *Regione Piemonte, Torino 2006, Le olimpiadi del territorio piemontese*. Regione Piemonte.

Reiche, D. (2017). *Success and failure of countries at the Olympic Games*. Routledge.

Ritchie, J. R. (2000). Turning 16 days into 16 years through Olympic legacies. *Event Management*, *6*(3), 155–165. London: Routledge.

Ritchie, J. R. B., & Lyons, M. (1990). Olympulse VI: A post-event assessment of resident reaction to the XV Olympic Winter Games. *Journal of Travel Research*, *28*(3), 14–23.

Ritchie, J. R. B., & Smith, B. H. (1991). The impact of a mega-event on host regional awareness: A longitudinal study. *Journal of Travel Research*, *30*(1), 3–10.

Roche, M. (1992). Mega-events and micro-modernisation: On the sociology of the new urban tourism. *British Journal of Sociology*, *43*, 563–600.

Roche, M. (2000). *Megaevents and modernity: Olympics and expos in the growth of global C*. Routledge. Available at: <https://citeseerx.ist.psu.edu/document?re pid=rep1&type=pdf&doi=1bbe07c9a1b53747a55c8db4d220a4072428faf8>.

Roche, M. (2002). Olympic and sport mega-events as media-events: Reflections on the globalisation paradigm. *Symposium a Quarterly Journal in Modern Foreign Literatures*, 1–12.

Roche, M. (2003). The Olympics and the development of "Global Society". In M. De Moragas, C. Kennett, & N. Puig (a cura di), *The legacy of the Olympic Games*. Document of the Olympic Museum. Losanna: International Olympic Committee.

Roche, M. (2006). Mega-events and modernity revisited: Globalization and the case of the Olympics. *The Sociological Review*, *54*(2_suppl), 27–40.

Rose, A., & Spiegel, M. (2009). *The Olympic effect*. Cambridge, MA: National Bureau of Economic Research.

Rossi, A. (1995). *L'architettura della città*. Città Studi Edizione.

Sainsbury T. (2012). Olympic Villages. In Gold J., M. M. G. (Eds), *Olympic Cities: City Agendas, Planning, and the World's Games, 1896* (3rd ed.). Routledge.

Salt Lake City Organizing Committee. (2002). *Official report of the XIX Olympic Winter Games Salt Lake 2002: 8–24 February 2002*.

Salzano, E. (2002). *Fondamenti di Urbanistica*. Edizioni Laterza.

Samonà, G. (1971). *L'urbanistica e L'avvenire della Città*. Bari: Laterza.

Sanders, B. (1995). *Briefing Document on Remediation and Ecological Investigations, Olympic Co-ordination Agency*. Homebush Bay: NSW.

Sassen, S. (1991). *The global city: New York, London, Tokyo*. Princeton, NJ: Princeton University Press.

Sassen, S. (2018). *Cities in a world economy* (5th edn). SAGE Publications.

Savitch, H. V. (1988). *Post-industrial cities, politics and planning in New York, Paris and London*. Princeton University Press.

Scamuzzi, S. (2002). Perché le città hanno bisogno di marketing ma solo alcune lo fanno con successo?. In L. Bobbio & C. Guala (a cura di), *Olimpiadi e grandi eventi. Verso Torino 2006* (pp. 87–93). Roma: Carocci.

Scherer, J. (2011). Olympic Villages and large-scale urban development: Crises of capitalism, deficits of democracy? *Sociology*, *45*(5), 782–797.

Scott, D., Steiger, R., Rutty, M., & Johnson, P. (2015). The future of the Olympic Winter Games in an era of climate change. *Current Issues in Tourism, 18*(10), 913–930.

Segre, A., & Scamuzzi, S. (2004). *Aspettando le Olimpiadi: Torino 2006: primo rapporto sui territori olimpici.*

Sennett, R. (1987). The fall of public man. *Capital & Class, 11*(1), 132–134.

Seoul Olympic Organizing Committee. (1989). *Official report: Games of the XXIVth Olympiad Seoul 1988.*

Shaw, C. A. (2012). The economics and marketing of the Olympic Games from bid phase to aftermath. In *The Palgrave handbook of Olympic studies* (pp. 248–260). Palgrave Macmillan UK.

Short, J. R. (2008). Globalization, cities and the Summer Olympics. *City, 12*(3), 321–340.

Smith, A. (2009). Theorising the relationship between major sport events and social sustainability. *Journal of Sport and Tourism, 14*(2–3), 109–120.

Smith, A. (2012). *Events and urban regeneration.* London: Routledge.

Smith, A. (2014). "De-Risking" east London: Olympic regeneration planning 2000–2012. *European Planning Studies, 22*(9), 1919–1939.

Smith, C. J., & Himmelfarb, K. M. G. (2007). Restructuring Beijing's social space: Observations on the Olympic Games in 2008. *Eurasian Geography and Economics, 48*(5), 543–554.

Smith, M. (2008). *When the games come to town: Host cities and the local impacts of the Olympics. London East Research Institute Working Papers, University of East London, London,* (December), pp. 1–95.

Smith, N. (1996). *The new urban frontier: Gentrification and the revanchist city.* Routledge.

Soja, E. (2000). *Postmetropolis: Critical studies of cities and regions|Wiley.* Wiley-Blackwell.

SOOC [Sochi Olympic Organising Committee]. (2009). Games 2014 will double Sochi power supply. Available at: <http://sochi2014.com/87868>.

SOOC. (2014). *Sochi 2014: Official Report.* Moscow: Organizing Committee of XXII Olympic Winter Games and XI Paralympic Winter Games of 2014.

Sordet P. (1996). The Olympic Village of Albertville' 1996. In *Hundred years of urban planning and shared experiences.* International Symposium on Olympic Villages. IOC.

Sorkin, M. (1992). *Variations on a Theme Park.* New York: Hill and Wang.

Spaziosport, G. B.-, Seoul, special number dedicated to the, & 1988, undefined. (n.d.). Architecture and the Games.

Spilling, O. (1998). Beyond Intermezzo? On the long-term industrial impacts of mega-events: The case of Lillehammer 1994. *Festival Management & Events Tourism*, *5*, 101–122.

Spilling, O. R. (1996). Mega event as strategy for regional development the case of the 1994 Lillehammer Winter Olympics. *Entrepreneurship and Regional Development*, *8*(4), 321–344.

Spooner, E. D., Morphett, D., Watt, M. E., Grunwald, G., & Zacharias, P. (2000). Solar Olympic Village case study. *Energy Policy*, *28*(14), 1059–1068.

Swyngedouw, E. (2004). Globalisation or "glocalisation"? Networks, territories and rescaling. *Cambridge Review of International Affairs*, *17*(1), 25–48.

Swyngedouw, E., Moulaert, F., & Rodriguez, A. (2002). Neoliberal urbanization in Europe: Large-scale urban development projects and the new urban policy. *Antipode*, *34*(3), 542–577.

Sydney Organizing Committee for the Olympic Games. (2001). *Official report of the XXVII Olympiad: Sydney 2000 Olympic games.*

Tanaka, Y. (1997). ITS Traffic management for Nagano Olympic Winter Games in Japan. *Mobility for everyone.* 4th World Congress on intelligent transport systems. Berlin, 1–5.

Xth Olympiade Committee of the Games of Los Angeles. (1933). *The games of the Xth Olympiad Los Angeles 1932*: Official report.

Terret, T. (2008). The Albertville Winter Olympics: Unexpected legacies – Failed expectations for regional economic development. *The International Journal of the History of Sport*, *25*(14), 1903–1921.

Theodoraki, E. (2007). *Olympic event organization* (1st edn). Routledge.

Tokyo 2020 Olympic Games Bid Committee. (2013). *Tokyo 2020: Discover tomorrow.*

Town and Country Planning Act. (1968). United Kingdom.

Turco, A. (1984). *Regione e regionalizzazione: Scritti di Roger Brunet.* Franco Angeli.

Turco, A. (1988). *Verso una teoria geografica della complessità.* Milano: Unicopli.

UK Government. (2011). *The Green Book: Appraisal and evaluation in central government – GOV.UK.*

UK Government. (2011). *The Magenta Book – GOV.UK.*

UN. (1980). *Patterns of urban and rural population growth.*

UN. (1992). *Agenda 21.* UN.

UN. (2005). *United Nations development program.* UN.

UN. (2016). *Culture: Urban future\Diversity of cultural expressions.*

UNEP. (2005). *World resources 2005: The wealth of the poor – Managing ecosystems to fight poverty.* Washington, DC: London Word Resources Institute.

UN-Habitat. (2009). *Planning sustainable cities: Global report on human settlements.*

UN-Habitat. (2015). *International guidelines on urban and territorial planning.*

United Nations. (2019). *The future is now.*

Urry, J. (1995). *Consuming places* (1st edn). Routledge Book.

VANOC (2009). *Olympic Games Impact (OGI) study for the 2010 Olympic and Paralympic Winter Games pre-games results report the Vancouver Organizing Committee for the 2010 Olympic and Paralympic games (VANOC).*

Vanolo, A. (2008). The image of the creative city: Some reflections on urban branding in Turin. *Cities, 25*(6), 370–382.

VanWynsberghe, R. (2015). The Olympic Games Impact (OGI) study for the 2010 Winter Olympic Games: Strategies for evaluating sport mega-events' contribution to sustainability. *International Journal of Sport Policy, 7*(1), 1–18.

VanWynsberghe, R., & Khan, S. (2007). Redefining case study. *International Journal of Qualitative Methods, 6*(2), 80–94.

VanWynsberghe, R., Derom, I., & Maurer, E. (2012). Social leveraging of the 2010 Olympic Games: "sustainability" in a city of Vancouver initiative. *Journal of Policy Research in Tourism, Leisure and Events, 4*(2), 185–205.

Venturi, M. (1994). *Grandi Eventi, La festivalizzazione della politica urbana*. Il Cardo.

Venturi, R. (1966). *Complexity and contradiction in Architecture*. MOMA.

Viehoff, V., & Poynter, G. (2018). *Mega-event cities: Urban legacies of global sports events* (1st edn). Routledge.

Ward, S. (1998). *Selling places: The marketing and promotion of towns and cities 1850–2000*. Routledge.

Weicker, F., & Company Management Consultants. (2003). *Vancouver agreement community assessment of 2010 Olympic Winter Games and Paralympic Games on Vancouver's inner-city neighbourhoods final report February 2003 prepared for the Vancouver agreement in conjunction with the Vancouver 2010 bid corporation.*

Westerbeek, H. M., Turner, P., & Ingerson, L. (2002). Key success factors in bidding for hallmark sporting events. *International Marketing Review, 19*(3), 303–322.

Wimmer, W. (1976). *Olympic buildings*. Edition: Leipzig.

Zervas, K. (2012). Anti-Olympic campaigns. In *The Palgrave handbook of Olympic studies* (pp. 533–548). Palgrave Macmillan UK.

Zou, Y., Mason, R., & Zhong, R. (2015). Modeling the polycentric evolution of post-Olympic Beijing: An empirical analysis of land prices and development intensity. *Urban Geography, 36*(5), 735–756.

Zukin, S. (1991). *Landscapes of power: From Detroit to Disney World*. Berkeley, CA: University' of California Press.

Zukin, S. (1998). Urban lifestyles: Diversity and standardisation in spaces of consumption. *Urban Studies, 35*(5–6), 825–839.

Zukin, S., Kasinitz, P., & Chen, X. (2015). *Global cities, local streets*. Routledge.

Index

9 781803 742946 *